Reviews of
Environmental Contamination
and Toxicology

VOLUME 230

For further volumes:
http://www.springer.com/series/398

# Reviews of Environmental Contamination and Toxicology

## With Cumulative and Comprehensive Index Subjects Covered Volumes 221–230

Editor
David M. Whitacre

Editorial Board

Maria Fernanda, Cavieres, Valparaiso, Chile • Charles P. Gerba, Tucson, Arizona, USA
John Giesy, Saskatoon, Saskatchewan, Canada • O. Hutzinger, Bayreuth, Germany
James B. Knaak, Getzville, New York, USA
James T. Stevens, Winston-Salem, North Carolina, USA
Ronald S. Tjeerdema, Davis, California, USA • Pim de Voogt, Amsterdam, The Netherlands
George W. Ware, Tucson, Arizona, USA

Founding Editor
Francis A. Gunther

**VOLUME 230**

## Coordinating Board of Editors

Dr. David M. Whitacre, *Editor*
*Reviews of Environmental Contamination and Toxicology*

5115 Bunch Road
Summerfield, North Carolina 27358, USA
(336) 634-2131 (PHONE and FAX)
E-mail: dmwhitacre@triad.rr.com

Dr. Erin R. Bennett, *Editor*
*Bulletin of Environmental Contamination and Toxicology*
*Great Lakes Institute for Environmental Research*

University of Windsor
Windsor, Ontario, Canada
E-mail: ebennett@uwindsor.ca

Peter S. Ross, *Editor*
*Archives of Environmental Contamination and Toxicology*

Fisheries and Oceans Canada
Institute of Ocean Sciences Sidney
British Colombia, Canada
E-mail: peter.s.ross@dfo-mpo.gc.ca

ISSN 0179-5953  ISSN 2197-6554 (electronic)
ISBN 978-3-319-04410-1  ISBN 978-3-319-04411-8 (eBook)
DOI 10.1007/978-3-319-04411-8
Springer Cham Heidelberg New York Dordrecht London

© Springer International Publishing Switzerland 2014
This work is subject to copyright. All rights are reserved by the Publisher, whether the whole or part of the material is concerned, specifically the rights of translation, reprinting, reuse of illustrations, recitation, broadcasting, reproduction on microfilms or in any other physical way, and transmission or information storage and retrieval, electronic adaptation, computer software, or by similar or dissimilar methodology now known or hereafter developed. Exempted from this legal reservation are brief excerpts in connection with reviews or scholarly analysis or material supplied specifically for the purpose of being entered and executed on a computer system, for exclusive use by the purchaser of the work. Duplication of this publication or parts thereof is permitted only under the provisions of the Copyright Law of the Publisher's location, in its current version, and permission for use must always be obtained from Springer. Permissions for use may be obtained through RightsLink at the Copyright Clearance Center. Violations are liable to prosecution under the respective Copyright Law.
The use of general descriptive names, registered names, trademarks, service marks, etc. in this publication does not imply, even in the absence of a specific statement, that such names are exempt from the relevant protective laws and regulations and therefore free for general use.
While the advice and information in this book are believed to be true and accurate at the date of publication, neither the authors nor the editors nor the publisher can accept any legal responsibility for any errors or omissions that may be made. The publisher makes no warranty, express or implied, with respect to the material contained herein.

Printed on acid-free paper

Springer is part of Springer Science+Business Media (www.springer.com)

# Foreword

International concern in scientific, industrial, and governmental communities over traces of xenobiotics in foods and in both abiotic and biotic environments has justified the present triumvirate of specialized publications in this field: comprehensive reviews, rapidly published research papers and progress reports, and archival documentations. These three international publications are integrated and scheduled to provide the coherency essential for nonduplicative and current progress in a field as dynamic and complex as environmental contamination and toxicology. This series is reserved exclusively for the diversified literature on "toxic" chemicals in our food, our feeds, our homes, recreational and working surroundings, our domestic animals, our wildlife, and ourselves. Tremendous efforts worldwide have been mobilized to evaluate the nature, presence, magnitude, fate, and toxicology of the chemicals loosed upon the Earth. Among the sequelae of this broad new emphasis is an undeniable need for an articulated set of authoritative publications, where one can find the latest important world literature produced by these emerging areas of science together with documentation of pertinent ancillary legislation.

Research directors and legislative or administrative advisers do not have the time to scan the escalating number of technical publications that may contain articles important to current responsibility. Rather, these individuals need the background provided by detailed reviews and the assurance that the latest information is made available to them, all with minimal literature searching. Similarly, the scientist assigned or attracted to a new problem is required to glean all literature pertinent to the task, to publish new developments or important new experimental details quickly, to inform others of findings that might alter their own efforts, and eventually to publish all his/her supporting data and conclusions for archival purposes.

In the fields of environmental contamination and toxicology, the sum of these concerns and responsibilities is decisively addressed by the uniform, encompassing, and timely publication format of the Springer triumvirate:

*Reviews of Environmental Contamination and Toxicology* [Vol. 1 through 97 (1962–1986) as Residue Reviews] for detailed review articles concerned with any aspects of chemical contaminants, including pesticides, in the total environment with toxicological considerations and consequences.

*Bulletin of Environmental Contamination and Toxicology* (Vol. 1 in 1966) for rapid publication of short reports of significant advances and discoveries in the fields of air, soil, water, and food contamination and pollution as well as methodology and other disciplines concerned with the introduction, presence, and effects of toxicants in the total environment.

*Archives of Environmental Contamination and Toxicology* (Vol. 1 in 1973) for important complete articles emphasizing and describing original experimental or theoretical research work pertaining to the scientific aspects of chemical contaminants in the environment.

Manuscripts for Reviews and the Archives are in identical formats and are peer reviewed by scientists in the field for adequacy and value; manuscripts for the Bulletin are also reviewed, but are published by photo-offset from camera-ready copy to provide the latest results with minimum delay. The individual editors of these three publications comprise the joint Coordinating Board of Editors with referral within the board of manuscripts submitted to one publication but deemed by major emphasis or length more suitable for one of the others.

Coordinating Board of Editors

# Preface

The role of Reviews is to publish detailed scientific review articles on all aspects of environmental contamination and associated toxicological consequences. Such articles facilitate the often complex task of accessing and interpreting cogent scientific data within the confines of one or more closely related research fields.

In the nearly 50 years since *Reviews of Environmental Contamination and Toxicology* ( formerly *Residue Reviews*) was first published, the number, scope, and complexity of environmental pollution incidents have grown unabated. During this entire period, the emphasis has been on publishing articles that address the presence and toxicity of environmental contaminants. New research is published each year on a myriad of environmental pollution issues facing people worldwide. This fact, and the routine discovery and reporting of new environmental contamination cases, creates an increasingly important function for *Reviews*.

The staggering volume of scientific literature demands remedy by which data can be synthesized and made available to readers in an abridged form. *Reviews* addresses this need and provides detailed reviews worldwide to key scientists and science or policy administrators, whether employed by government, universities, or the private sector.

There is a panoply of environmental issues and concerns on which many scientists have focused their research in past years. The scope of this list is quite broad, encompassing environmental events globally that affect marine and terrestrial ecosystems; biotic and abiotic environments; impacts on plants, humans, and wildlife; and pollutants, both chemical and radioactive; as well as the ravages of environmental disease in virtually all environmental media (soil, water, air). New or enhanced safety and environmental concerns have emerged in the last decade to be added to incidents covered by the media, studied by scientists, and addressed by governmental and private institutions. Among these are events so striking that they are creating a paradigm shift. Two in particular are at the center of everincreasing media as well as scientific attention: bioterrorism and global warming. Unfortunately, these very worrisome issues are now superimposed on the already extensive list of ongoing environmental challenges.

The ultimate role of publishing scientific research is to enhance understanding of the environment in ways that allow the public to be better informed. The term "informed public" as used by Thomas Jefferson in the age of enlightenment conveyed the thought of soundness and good judgment. In the modern sense, being "well informed" has the narrower meaning of having access to sufficient information. Because the public still gets most of its information on science and technology from TV news and reports, the role for scientists as interpreters and brokers of scientific information to the public will grow rather than diminish. Environmentalism is the newest global political force, resulting in the emergence of multinational consortia to control pollution and the evolution of the environmental ethic. Will the new politics of the twenty-first century involve a consortium of technologists and environmentalists, or a progressive confrontation? These matters are of genuine concern to governmental agencies and legislative bodies around the world.

For those who make the decisions about how our planet is managed, there is an ongoing need for continual surveillance and intelligent controls to avoid endangering the environment, public health, and wildlife. Ensuring safety-in-use of the many chemicals involved in our highly industrialized culture is a dynamic challenge, for the old, established materials are continually being displaced by newly developed molecules more acceptable to federal and state regulatory agencies, public health officials, and environmentalists.

*Reviews* publishes synoptic articles designed to treat the presence, fate, and, if possible, the safety of xenobiotics in any segment of the environment. These reviews can be either general or specific, but properly lie in the domains of analytical chemistry and its methodology, biochemistry, human and animal medicine, legislation, pharmacology, physiology, toxicology, and regulation. Certain affairs in food technology concerned specifically with pesticide and other food-additive problems may also be appropriate.

Because manuscripts are published in the order in which they are received in final form, it may seem that some important aspects have been neglected at times. However, these apparent omissions are recognized, and pertinent manuscripts are likely in preparation or planned. The field is so very large and the interests in it are so varied that the editor and the editorial board earnestly solicit authors and suggestions of underrepresented topics to make this international book series yet more useful and worthwhile.

Justification for the preparation of any review for this book series is that it deals with some aspect of the many real problems arising from the presence of foreign chemicals in our surroundings. Thus, manuscripts may encompass case studies from any country. Food additives, including pesticides, or their metabolites that may persist into human food and animal feeds are within this scope. Additionally, chemical contamination in any manner of air, water, soil, or plant or animal life is within these objectives and their purview.

Manuscripts are often contributed by invitation. However, nominations for new topics or topics in areas that are rapidly advancing are welcome. Preliminary communication with the editor is recommended before volunteered review manuscripts are submitted.

Summerfield, NC, USA                                                              David M. Whitacre

# Contents

**Removal of Vapor-Phase Elemental Mercury
from Stack Emissions with Sulfur-Impregnated Activated Carbon** .......... 1
Mohammad Hossein Sowlat, Mohammad Abdollahi,
Hamed Gharibi, Masud Yunesian, and Noushin Rastkari

**Setting Water Quality Criteria in China:
Approaches for Developing Species Sensitivity
Distributions for Metals and Metalloids** ....................... 35
Yuedan Liu, Fengchang Wu, Yunsong Mu, Chenglian Feng,
Yixiang Fang, Lulu Chen, and John P. Giesy

**Toxicity Reference Values for Protecting Aquatic Birds
in China from the Effects of Polychlorinated Biphenyls** ........................ 59
Hailei Su, Fengchang Wu, Ruiqing Zhang, Xiaoli Zhao,
Yunsong Mu, Chenglian Feng, and John P. Giesy

**Fabricated Nanoparticles: Current Status
and Potential Phytotoxic Threats** ....................... 83
Tushar Yadav, Alka A. Mungray, and Arvind K. Mungray

**Status of Heavy Metal Residues in Fish Species of Pakistan** ...................... 111
Majid Hussain, Said Muhammad, Riffat N. Malik,
Muhammad U. Khan, and Umar Farooq

**Index**............................................................................................................. 133

# Removal of Vapor-Phase Elemental Mercury from Stack Emissions with Sulfur-Impregnated Activated Carbon

Mohammad Hossein Sowlat, Mohammad Abdollahi, Hamed Gharibi, Masud Yunesian, and Noushin Rastkari

## Contents

| | | |
|---|---|---|
| 1 | Introduction | 2 |
| 2 | Review Approach and Methodology | 3 |
| 3 | S-Impregnated AC for Removing Hg | 4 |
| | 3.1 The Removal Efficiency of Sulfur-Impregnated ACs vs. Virgin ACs | 26 |
| | 3.2 Effect of Operational Temperature | 27 |
| | 3.3 Effect of Inlet $Hg^0$ Concentration | 28 |
| | 3.4 Effect of Sulfur-to-Carbon (S/C) Ratio and Sulfur Content | 28 |
| | 3.5 Effect of Impregnation Temperature and Time | 29 |
| 4 | Summary | 31 |
| References | | 32 |

M.H. Sowlat • N. Rastkari (✉)
Center for Air Pollution Research (CAPR), Institute for Environmental Research (IER),
Tehran University of Medical Sciences, Kargar Shomali St., Tehran, Iran
e-mail: nr_raskari@yahoo.com

M. Abdollahi
Faculty of Pharmacy, and Pharmaceutical Sciences Research Center, and Endocrinology
& Metabolism Research Center, Tehran University of Medical Sciences,
Enqelab Sq., Tehran, Iran

H. Gharibi
School of Public Health, Shahroud University of Medical Sciences, Shahroud, Iran

M. Yunesian
Center for Air Pollution Research (CAPR), Institute for Environmental Research (IER),
Tehran University of Medical Sciences, Kargar Shomali St., Tehran, Iran

Department of Environmental Health Engineering, School of Public Health,
Tehran University of Medical Sciences, Enqelab Sq., Tehran, Iran

# 1 Introduction

Mercury (Hg) is a trace element that can cause severe health effects in exposed individuals. Clakson (1993) reported that the high solubility of mercury vapor in cell membranes results in its rapid absorption and transport to target tissues, including kidneys and brain. After wet or dry deposition on vegetation, Hg may reach the human body via food ingestion, where it is transported to target tissues and induces adverse effects (Berlin 1979). Hg is listed as a hazardous air pollutant (HAP) in the US Clean Air Act Amendments (CAAA) of 1990, mainly because of its severe public health and environmental affects. Approximately 2,500 t of Hg are emitted annually from natural sources, and another 3,500 t are emitted via anthropogenic sources, such as coal-fired power plants and solid waste combustors (Johnson 1997; Pacyna and Munch 1991). In the USA alone, 150 t of Hg is annually released from anthropogenic sources (Brenneman et al. 2000).

Hg is present in the stack emissions of coal-fired power plants and solid waste combustors in three primary forms: elemental Hg ($Hg^0$); the oxidized form, e.g., mercuric chloride ($HgCl_2$); and as particulate-bound Hg ($Hg_p$) (Galbreath and Zygarlicke 1996; Pavlish et al. 2003). Although $Hg_p$ can be captured by electrostatic precipitations (ESP) and the oxidized Hg form can be removed by flue gas desulfurization (FGD) or fabric filters (also called baghouse) (Galbreath and Zygarlicke 1996; Pavlish et al. 2003; Volland 1991), it is much more difficult to remove $Hg^0$ from stack emissions because of its low solubility in water, high equilibrium vapor pressure, and low melting point (Schroedor et al. 1991; Schuster 1991; Weast 1983).

Therefore, other methods have been proposed for controlling $Hg^0$ emissions, among which granular activated carbon (GAC) and powdered activated carbon (PAC) have shown promising results (Sinha and Walker 1972; Young et al. 1994). Despite their substantially high efficiency for removing $Hg^0$, activated carbons (ACs) have high operating costs (Cal et al. 2000; Dorman et al. 2002), necessitating remarkable improvements in their performance if they are to be economically viable. Early studies revealed that impregnation of ACs with sulfur significantly increased the performance for removing $Hg^0$ from stack emissions (Sinha and Walker 1972). More recently, many studies have been conducted to explore the effect of different parameters on the performance of sulfur-impregnated ACs for removing $Hg^0$ and to compare their efficiency with virgin forms of AC. Notwithstanding, after performing a search of the literature, no systematic review was found that addressed the relative efficiency of or effect of operational and impregnation parameters on virgin and sulfur-impregnated ACs for removing $Hg^0$.

Therefore, our main objective in this chapter is to systematically review the existing data and evidence relating to the relative efficiency of sulfur-impregnated ACs, vs. virgin ACs, for removing $Hg^0$ from sources of industrial stack emissions. A second goal is to present an overview of the effect of different operational and impregnation parameters on removal efficiency of virgin and sulfur-impregnated ACs.

In preparing this chapter, we relied on the methods of Khan et al. (2003), which emphasize the importance of a systematic approach in preparing reviews. Utilizing such rigorous methods is designed to enhance the validity, trustworthiness, and strength of the conclusions presented (Green et al. 2001).

## 2 Review Approach and Methodology

Khan et al. (2003) proposed the following four major steps for conducting an appropriate systematic review: (1) formulate the framing question (or study question), (2) identify the literature that is relevant to the topic by selecting appropriate bibliographic databases and search terms and by defining inclusion and exclusion criteria, and (3) assess the methodological quality (with respect to sample size, control group, instrumentation, and ranges selected for the key variables) of the selected papers. Finally, extract and summarize the relevant findings of each study. These steps are summarized in Fig. 1.

*Step 1—Framing Question*
The main research question formulated for the present systematic review was the following: "What is the efficiency of sulfur-impregnated ACs compared to virgin ones for the removal of vapor-phase elemental Hg from stack emissions?" A minor question was also posed to address the following point: "What are the effects of operational parameters, such as operating temperature, impregnation temperature, or inlet $Hg^0$ concentration, on the adsorption capacity of sulfur-impregnated ACs?"

*Step 2—Relevant Literature*
We searched "Web of Science" and "Scopus" databases, mainly because they cover the great majority of cogent literature on our topic (viz., Elsevier, Springer, American Chemical Society (ACS), Taylor and Francis, and Wiley). We formulated the search strategy by employing a combination of the following: search terms, including "sulfur impregnation," "activated carbon," "elemental mercury," "vapor phase," and all

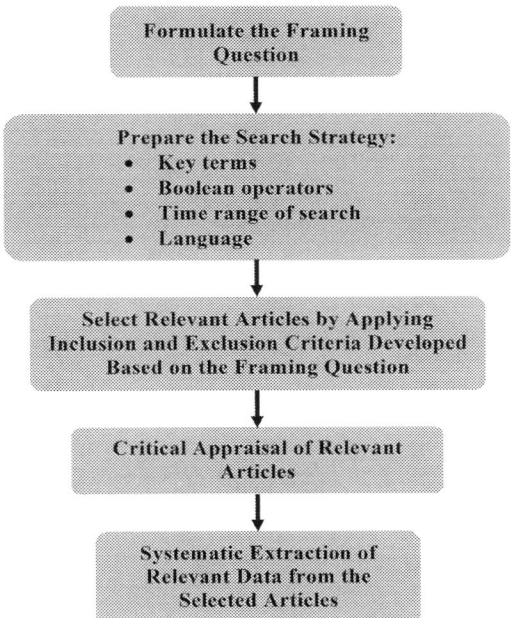

**Fig. 1** Process steps used to prepare this systematic review

of their possible variations and synonyms, and Boolean operators, such as "AND," "OR," and "NOT." The final terms used in the search strategy were:

((mercury OR Hg OR "mercury vapor" OR "Hg vapor" OR "elemental mercury" OR "elemental Hg" OR "vapor phase mercury" OR "vapor phase Hg" OR "mercury emission" OR "Hg emission") AND (capture OR sorption OR adsorption OR "adsorption capacity" OR removal OR uptake OR "uptake capacity" OR "scavenging capacity")) AND (((("gaseous sulfur" OR $SO_2$ OR "sulfur dioxide" OR $H_2S$ OR "hydrogen sulfide" OR sulfur OR "elemental sulfur" OR "organic sulfur") AND (impregnat* OR deposit* OR chemisor* OR incorporat* OR fixat*)) OR sulfuri?ed) AND ("activated carbon" OR AC OR "carbon sorbent" OR "granular activated carbon" OR GAC))

These databases were searched up to March 2012 using the above search strategy. To ensure that relevant papers were not missed, we reviewed the reference list of the retrieved papers for additional potentially relevant studies.

We defined inclusion criteria as being all original papers published in English that reported on the efficiency of sulfur-impregnated ACs vs. virgin ones for removing gas-phase $Hg^0$ from stack emissions. We excluded papers published in any language other than English, those reporting findings on any phase other than gas (such as liquid phase), studies conducted on ACs impregnated with other chemicals (halogens, for example), or those published on any other forms of Hg (e.g., oxidized forms). We also excluded studies in which sulfur impregnation of ACs was done without subsequent testing for $Hg^0$ removal or those not comparing the $Hg^0$ adsorption capacity of sulfur-impregnated ACs with virgin ones.

*Step 3—Quality Assessment*

As stated earlier, we did not find any systematic reviews that addressed sulfur-impregnated AC efficiency for removing gas-phase $Hg^0$. Therefore, we developed a checklist of five questions that allowed us to rate the quality of each paper we included in this review. Each question in the checklist was allocated 1 point (Yes=1 point, No=0 point); thus, the overall quality scale ranged between 1 and 5 points, with studies scoring 2 or less being rated as low quality, whereas those with 3 or higher being rated as high quality. We excluded all low-quality studies from our review. Finally, after assessing the quality of each included paper, information was extracted on impregnation parameters (i.e., S/C—sulfur to carbon—ratio, impregnation temperature, impregnation time, and sulfur type), carbon characteristics (i.e., surface area, sulfur content, and pore volume) before and after the impregnation, operational settings (i.e., inlet $Hg^0$ concentration and bed temperature), and outcomes.

# 3 S-Impregnated AC for Removing Hg

Figure 2 depicts the flow chart of the process used for selecting and evaluating studies to include in this review. As shown in this figure, our search strategy yielded a total of 1,566 hits: 70 from Web of Science and 1,496 from Scopus. In stage 1, duplications (i.e., the same articles found in both databases) were removed; then, potentially relevant articles were screened against eligibility criteria, and their full texts were retrieved ($n=54$). In stage 2, to ensure that no relevant papers were missed, the

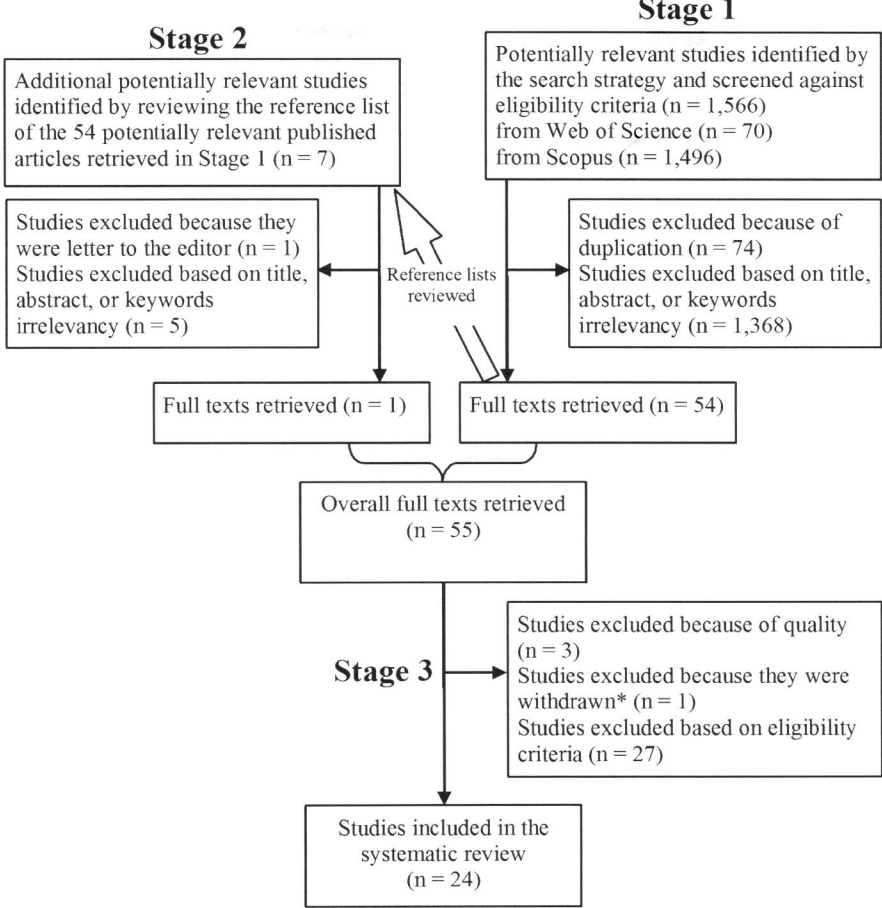

**Fig. 2** Flow chart of the process used for selecting and evaluating studies to include in this review

reference lists of these 54 articles were reviewed and 7 additional potentially relevant papers were found. Of these, five papers were excluded because of irrelevancy and one was excluded because, rather than being a paper, it was a letter to the editor. Therefore, based on eligibility criteria, only one paper was added to the list of potentially relevant papers, and its full text was retrieved. This yielded a total of 55 potentially relevant studies. In stage 3, the full texts of these 55 papers were reviewed, of which 30 were excluded from inadequate (low) quality or from not meeting eligibility criteria. One paper was excluded because it had been withdrawn by the Editor in Chief of the target journal. The reason for the exclusion was because the authors had failed to comply with the ethical criteria applied by the publisher. This left a total of 24 articles that were included in the review. A summary of the included studies, in terms of carbon characteristics, impregnation setting, operational setting, outcome measures, and main findings is presented in Table 1.

**Table 1** Summary findings of the papers included in this systematic review

| Study | Virgin carbons | | | Impregnation settings | | | |
| --- | --- | --- | --- | --- | --- | --- | --- |
| | Surface area (m²/g) | S% (by wt) | Pore volume (cm³/g) | Time (h) | Temperature (°C) | S/C ratio | Sulfur type |
| Otani et al. (1988) | 1,250 | 0 | 0.56 | – | – | – | CS$_2$ |
| Krishnan et al. (1994) | 547–964 | <1 | – | – | – | – | – |
| Vidic (1996) | – | 0.76 | – | – | 600 | – | – |

| Impregnated carbons | | | | | | |
|---|---|---|---|---|---|---|
| Surface area (m²/g) | S% (by wt) | Pore volume | Operation temperature (°C) | Hg⁰ input (μg/m³) | Outcome measure | Main findings |
| 710 (at 13.1%) | 0–13.1 | – | 36 | 6.42 mg/m³ | Breakthrough curves; Hg⁰ adsorption capacity | Impregnation increased the S content of ACs but decreased their surface area; higher S content led to increased Hg⁰ adsorption capacities; therefore, impregnated ACs had higher capacities than did virgin ACs |
| 715–1,078 | 7 | – | 23, 140 | 30, 60 ppb | Hg⁰ adsorption capacity | At higher temperature, sulfur-impregnated ACs performed better than virgin ones; the adsorption capacity of virgin ACs decreased with increasing temperatures, but sulfur-impregnated ACs increased with temperature increases; among virgin ACs with the same S content, the one with higher surface area had higher Hg⁰ adsorption capacity; increased Hg⁰ input significantly enhanced the efficiency of sulfur-impregnated ACs at all temperatures, with this effect being less significant for virgin ACs |
| – | 7.61–9.24 | – | 25–140 | 25–115 | Breakthrough curves; Hg⁰ adsorption capacity | The Hg⁰ adsorption capacity of virgin ACs decreased with increasing temperatures, while that of sulfur-impregnated ACs increased with temperature; sulfur-impregnated ACs performed better than virgin ones if impregnated with about the same S content and sulfur distribution |

(continued)

**Table 1** (continued)

| | Virgin carbons | | | Impregnation settings | | | |
|---|---|---|---|---|---|---|---|
| Study | Surface area ($m^2/g$) | S% (by wt) | Pore volume ($cm^3/g$) | Time (h) | Temperature (°C) | S/C ratio | Sulfur type |
| Vidic and McLaughlin (1996) | – | 0.76 | – | – | 600 | – | – |
| Korpiel and Vidic (1997) | 1,026 | 0.7 | – | – | 600, Approx. 200 | – | – |
| Hsi et al. (1998) | 353–994 | 0.8–2.6 | 0.2–0.7 | – | 600 | – | Elemental |

| Impregnated carbons | | | | | | |
|---|---|---|---|---|---|---|
| Surface area (m$^2$/g) | S% (by wt) | Pore volume | Operation temperature (°C) | Hg$^0$ input (μg/m$^3$) | Outcome measure | Main findings |
| – | 7.61–9.24 | – | 25–140 | 25–115 | Breakthrough curves; Hg$^0$ adsorption capacity | For virgin ACs, the Hg$^0$ adsorption capacity increased with increasing Hg$^0$ inputs and decreasing temperatures, whereas that of sulfur-impregnated ACs increased with temperature; sulfur-impregnated ACs were more effective than virgin ACs in removing Hg$^0$ |
| 482–824 | 9.7–10.0 | – | 25, 90, 150 | 55, 684 | Breakthrough curves; Hg$^0$ adsorption capacity | Higher impregnation temperatures led to more even distribution of sulfur in the pore structure and stronger bonding of sulfur to carbon; therefore, for ACs impregnated at 600 °C, the Hg adsorption did not deteriorate even at operating temperatures as high as 140 °C; however, generally, both types of sulfur-impregnated ACs had higher efficiencies at higher temperatures; higher Hg$^0$ inputs increased the removal efficiency of ACs impregnated at 600 °C, although efficiency for ACs impregnated at about 200 °C decreased |
| 532–670 | 11.5–11.9 | – | 49–82 | 45–68 | Hg$^0$ adsorption capacity | For ACs with comparable surface areas, the Hg$^0$ adsorption capacity of those made from high-sulfur coals was higher than that from low-sulfur coals; for ACs from high-sulfur coals, the Hg$^0$ adsorption capacity increased with increasing sulfur content; for both types, the Hg$^0$ adsorption capacity increased with surface area; sulfur impregnation markedly increased the sorption capacity of the ACs from low-sulfur coals while not significantly changing that of the ACs from high-sulfur coals |

(continued)

**Table 1** (continued)

| Study | Virgin carbons | | | Impregnation settings | | | |
| --- | --- | --- | --- | --- | --- | --- | --- |
| | Surface area (m$^2$/g) | S% (by wt) | Pore volume (cm$^3$/g) | Time (h) | Temperature (°C) | S/C ratio | Sulfur type |
| Liu et al. (1998) | 987.7–1,026.0 | 0.51–0.73 | – | 2 | Approx. 200, 250, 400, 600 | 1/2–4/1 | Elemental |
| Vidic et al. (1998) | 1,026 | 0.7 | – | – | Approx. 200, 600 | – | – |

| Impregnated carbons | | | | | | |
|---|---|---|---|---|---|---|
| Surface area (m$^2$/g) | S% (by wt) | Pore volume | Operation temperature (°C) | Hg$^0$ input (μg/m$^3$) | Outcome measure | Main findings |
| 164.4–909.5 | 7.11–38.5 | – | 140 | 55 | Hg$^0$ adsorption capacity | Hg$^0$ adsorption capacity increased with increasing impregnation temperatures and S/C ratios (up to a specific ratio of 2/1); higher impregnation temperatures led to the formation of ACs with larger surface areas, while higher S/C ratios decreased their surface area; higher impregnation temperatures also decreased the sulfur content of the impregnated ACs; sulfur-impregnated ACs exhibited significantly higher adsorption capacities compared to virgin and commercially available sulfur-impregnated ACs |
| 482–824 | 9.7–10.0 | – | 25, 140 | 110, 380, 1,080 | Hg adsorption capacity | For virgin ACs, an increase in Hg$^0$ inlet increased the Hg$^0$ adsorption capacity; also, the sorption capacity decreased with increasing operating temperatures, especially at higher Hg$^0$ inlets; for sulfur-impregnated ACs, the Hg$^0$ adsorption capacity increased with an increase in Hg$^0$ inlet, while operating temperature did not exhibit any significant impact; for the commercially available ACs impregnated at about 150–200 °C, an increase of operating temperature and inlet Hg$^0$ decreased the Hg$^0$ adsorption capacity; both sulfur-impregnated ACs had significantly higher Hg$^0$ adsorption capacities than virgin ACs, particularly at lower temperatures |

(continued)

**Table 1** (continued)

| Study | Virgin carbons | | | Impregnation settings | | | |
|---|---|---|---|---|---|---|---|
| | Surface area (m$^2$/g) | S% (by wt) | Pore volume (cm$^3$/g) | Time (h) | Temperature (°C) | S/C ratio | Sulfur type |
| Vitolo and Pini (1999) | – | – | – | – | – | – | – |
| Granite et al. (2000) | 650–900 | 0.4–0.9 | – | – | 600 | – | – |
| Karatza et al. (2000) | – | – | – | – | Approx. 200 | – | – |

| Impregnated carbons | | | Operation temperature (°C) | $Hg^0$ input (μg/m$^3$) | Outcome measure | Main findings |
|---|---|---|---|---|---|---|
| Surface area (m$^2$/g) | S% (by wt) | Pore volume | | | | |
| 460–503 | 12.4–23.8 | 163–188 mm$^3$/g | 25, 60, 70 | – | $Hg^0$ adsorption capacity | Higher sulfur content up to a specific point led to smaller surface areas; the $Hg^0$ adsorption capacity increased with increasing operating temperatures; the $Hg^0$ adsorption capacity also increased with increasing sulfur content up to a specific point; as the sulfur content increased, the fraction of sulfur reacting with $Hg^0$ to form HgS decreased |
| 690–790 | 5.9–7.6 | – | 138, 177 | – | Breakthrough curves; $Hg^0$ adsorption capacity | Sulfur-impregnated ACs indicated markedly higher $Hg^0$ adsorption capacities than virgin ones; sulfur-impregnated ACs performed better at higher operating temperatures; sulfur-impregnated ACs had significantly higher $Hg^0$ adsorption capacities than virgin ACs |
| 628 | Approx. 10.0 | – | 120, 150 | 2.24–3.93 mg/m$^3$ | Breakthrough curves | Inlet $Hg^0$ did not have any significant impact on the $Hg^0$ adsorption capacity at the concentration range evaluated; higher operating temperatures deteriorated the $Hg^0$ adsorption capacity of the sulfur-impregnated AC |

(continued)

**Table 1** (continued)

| Study | Virgin carbons | | | Impregnation settings | | | |
|---|---|---|---|---|---|---|---|
| | Surface area ($m^2/g$) | S% (by wt) | Pore volume ($cm^3/g$) | Time (h) | Temperature (°C) | S/C ratio | Sulfur type |
| Kwon and Vidic (2000) | 1,020 | 0.1 | – | 0.25, 0.05, 1, 2 | 200, 600 | – | Elemental, $H_2S$ |
| Liu et al. (2000) | 988–1,026 | 0.5–0.7 | – | 2 | Approx. 200, 600 | 1/5–4/1 | Elemental |

| Impregnated carbons | | | | | | |
|---|---|---|---|---|---|---|
| Surface area (m$^2$/g) | S% (by wt) | Pore volume | Operation temperature (°C) | Hg$^0$ input (μg/m$^3$) | Outcome measure | Main findings |
| <50–820 | 10.0–50.8 | – | 140 | 55 | Breakthrough curves; Hg$^0$ adsorption capacity | Sulfur-impregnated ACs performed much better than the virgin ACs for Hg$^0$ removal; Hg$^0$ adsorption capacity of the sulfur-impregnated ACs increased with increasing sulfur contents up to a specific point; at 200 °C, longer impregnation times remarkably increased the sulfur content of the sulfur-impregnated ACs, while significantly deteriorating their surface areas; hence, overall, the Hg$^0$ adsorption capacities of ACs impregnated at 200 °C decreased with increasing impregnation times; increasing of the impregnation temperature increased the Hg$^0$ adsorption capacity of the ACs because of more even distribution of sulfur in carbon matrix, though the sulfur content was lower at higher temperatures; the percentage of decrease in the surface area was also dramatically lower at higher impregnation temperatures than at lower temperatures |
| 789–905 | 7.9–12.9 | – | 140 | 55 | Hg$^0$ adsorption capacity | Sulfur-impregnated ACs performed much better than virgin ACs for Hg$^0$ removal; Hg$^0$ adsorption capacity increased with increasing S/C ratios (up to a specific ratio of 2/1) and, therefore, the final sulfur content; impregnation protocol did not have any significant effect on the sulfur content and surface area and, in turn, on the Hg$^0$ adsorption capacity of the ACs |

(continued)

**Table 1** (continued)

| Study | Virgin carbons ||| Impregnation settings ||||
|---|---|---|---|---|---|---|---|
| | Surface area (m²/g) | S% (by wt) | Pore volume (cm³/g) | Time (h) | Temperature (°C) | S/C ratio | Sulfur type |
| Hsi et al. (2001) | 1,971 | 0 | – | 6 | 250–650 | – | Elemental |
| Hsi et al. (2002) | 503–1,405 | 0–1.2 | 0.391–1.169 | 6 | 400 | 1/1 | Elemental |

| Impregnated carbons | | | | | | |
|---|---|---|---|---|---|---|
| Surface area ($m^2/g$) | S% (by wt) | Pore volume | Operation temperature (°C) | $Hg^0$ input ($\mu g/m^3$) | Outcome measure | Main findings |
| 4–1,816 | 6–64 | – | 135 | – | $Hg^0$ adsorption capacity | For impregnated ACs, an increase in the impregnation temperature significantly increased their surface area while decreasing their sulfur content; all sulfur-impregnated ACs performed better than the virgin ACs, though their surface areas were lower; the adsorption capacity of impregnated ACs increased with increasing impregnation temperatures up to 400 °C and decreased afterwards |
| 160–787 | 9.4–22.4 | 0.121–0.621 $cm^3/g$ | 163 | 50±20 | $Hg^0$ adsorption capacity | For sulfur-impregnated ACs, the higher the amount of sulfur added to them, the lower was their surface area; sulfur impregnation significantly increased the $Hg^0$ adsorption capacity of ACs; sulfur-impregnated ACs with higher sulfur content exhibited higher sorption capacities for $Hg^0$, although their surface areas were lower; for the concentration range studied, sulfur content was well correlated with $Hg^0$ adsorption capacity at a sulfur content of about 10% or higher ($R^2=0.86$); at a sulfur content <5%, surface area also became a significant parameter ($R^2=80$) |

(continued)

**Table 1** (continued)

| | Virgin carbons | | | Impregnation settings | | | |
|---|---|---|---|---|---|---|---|
| Study | Surface area ($m^2/g$) | S% (by wt) | Pore volume ($cm^3/g$) | Time (h) | Temperature (°C) | S/C ratio | Sulfur type |
| Lee and Park (2003) | 1,008–1,237 | Approx. 0 | – | – | Approx. 200, 400 | – | – |
| Ho et al. (2004) | 540 | <1 | – | – | – | – | – |

| Impregnated carbons | | | | | | |
|---|---|---|---|---|---|---|
| Surface area (m$^2$/g) | S% (by wt) | Pore volume | Operation temperature (°C) | Hg$^0$ input (μg/m$^3$) | Outcome measure | Main findings |
| 462–573 | 1–15 | – | 30, 70, 100, 140 | 160 | Hg$^0$ adsorption capacity | For virgin ACs with a sulfur content of approximately 0%, the highest Hg$^0$ adsorption capacity belonged to the one with the highest surface area and vice versa; the ACs impregnated at 400 °C had a significantly higher Hg$^0$ adsorption capacity than the commercially available sulfur-impregnated ACs, which are commonly activated at temperatures in the range of 150–200 °C; for sulfur-impregnated ACs, the Hg$^0$ adsorption capacity increased with an increase in the sulfur content; the Hg$^0$ adsorption capacity of the sulfur-impregnated AC also increased with increasing the operating temperature up to a specific point and decreased afterwards |
| 429.7 | Approx. 10 | – | 25 | 25.3 | Hg$^0$ adsorption capacity | Due to the low operating temperature, the virgin ACs performed better than the commercially available sulfur-impregnated ACs; under such conditions, the surface area and particle size of the activated carbons became the controlling factors |

(continued)

**Table 1** (continued)

| Study | Virgin carbons | | | Impregnation settings | | | |
|---|---|---|---|---|---|---|---|
| | Surface area (m$^2$/g) | S% (by wt) | Pore volume (cm$^3$/g) | Time (h) | Temperature (°C) | S/C ratio | Sulfur type |
| Yan et al. (2004) | 850–1,350 | 0 | – | – | – | – | – |
| Skodras et al. (2005) | 343–816 | 0.76–0.96 | 0.283–0.489 | 2 | 600 | – | Elemental |

| Impregnated carbons | | | Operation temperature (°C) | $Hg^0$ input ($\mu g/m^3$) | Outcome measure | Main findings |
|---|---|---|---|---|---|---|
| Surface area ($m^2/g$) | S% (by wt) | Pore volume | | | | |
| 950–1,100 | 10–15 | – | 30, 90, 140 | 8.1, 74.3 | $Hg^0$ adsorption capacity | Sulfur-impregnated ACs exhibited significantly higher $Hg^0$ adsorption capacities than the virgin ones; the $Hg^0$ adsorption capacity of the virgin AC deteriorated as the operating temperature increased; for sulfur-impregnated ACs, the $Hg^0$ adsorption capacity increased with increasing temperature up to a specific point and decreased afterwards; the sulfur-impregnated ACs managed to maintain high $Hg^0$ adsorption capacities as the inlet Hg concentration increased, but this change deteriorated that of virgin ACs |
| 310–779 | 4.37–7.16 | 0.265–0.447 $cm^3/g$ | 50, 100, 150, 200 | 0.35 $ng/cm^3$ | Breakthrough curves; $Hg^0$ adsorption capacity | Sulfur-impregnated ACs performed much better than their virgin counterparts; among impregnated ACs, the one with higher sulfur content, surface area, and pore volume exhibited higher $Hg^0$ adsorption capacity; the $Hg^0$ adsorption capacity of both virgin and sulfur-impregnated ACs decreased as the operating temperature increased, especially at temperatures higher than 100 °C |

(continued)

**Table 1** (continued)

| Study | Virgin carbons | | | Impregnation settings | | | |
|---|---|---|---|---|---|---|---|
| | Surface area (m$^2$/g) | S% (by wt) | Pore volume (cm$^3$/g) | Time (h) | Temperature (°C) | S/C ratio | Sulfur type |
| Feng et al. (2006a, b) | 920–1,950 | 0.2–0.75 | 0.374–0.806 | 2 | 200, 300, 400, 600, 800 | 1/1 | H$_2$S, Elemental |
| Feng et al. (2006c) | 920–1,950 | 0.2 | 0.371–0.741 | 2–24 | 80, 150 | – | H$_2$S |

| Impregnated carbons | | | Operation temperature (°C) | $Hg^0$ input ($\mu g/m^3$) | Outcome measure | Main findings |
|---|---|---|---|---|---|---|
| Surface area ($m^2/g$) | S% (by wt) | Pore volume | | | | |
| – | 2.9–7.9 | – | 140 | 350 | $Hg^0$ adsorption capacity | Virgin ACs with higher surface areas and pore volumes had higher $H_2S$ uptake than those with lower surface areas and pore volumes, so their final sulfur contents were higher; the sulfur content of the ACs also significantly increased with increasing impregnation temperature; the $Hg^0$ adsorption capacity of the sulfur-impregnated ACs dramatically increased with increasing impregnation temperatures up to a specific point (600 °C), but a significant decrease was observed afterwards; higher sulfur content also led to higher $Hg^0$ adsorption capacities for the sulfur-impregnated ACs |
| 8–1,880 | 4.1–30.5 | 0.005–0.714 $cm^3/g$ | 140 | 350 | $Hg^0$ adsorption capacity | Higher impregnation temperatures led to higher sulfur uptake, thus increasing the final sulfur content of the impregnated ACs; extended impregnation times increased the sulfur content of the ACs while significantly decreasing their surface areas and pore volumes; for virgin activated carbons, the $Hg^0$ adsorption capacity increased with increasing surface areas and pore volumes; sulfur-impregnated ACs performed much better than their virgin counterparts in removing $Hg^0$ from airstream; the $Hg^0$ adsorption capacity of the impregnated ACs increased with an increase in the sulfur content up to a specific point and decreased afterwards |

(continued)

**Table 1** (continued)

| | Virgin carbons | | | Impregnation settings | | | |
|---|---|---|---|---|---|---|---|
| Study | Surface area ($m^2/g$) | S% (by wt) | Pore volume ($cm^3/g$) | Time (h) | Temperature (°C) | S/C ratio | Sulfur type |
| Nabais et al. (2006) | 997–1,256 | | 0.37–0.49 | 1, 2, 3, 4 | 600, 800 | – | $H_2S$, Elemental |
| Lu et al. (2011) | 1,030 | – | 0.405 | 0.5–2.5 | 300, 350, 400, 500, 550, 600 | 1/1 | – |

| Impregnated carbons | | | | | | |
|---|---|---|---|---|---|---|
| Surface area ($m^2/g$) | S% (by wt) | Pore volume | Operation temperature (°C) | $Hg^0$ input ($\mu g/m^3$) | Outcome measure | Main findings |
| 848–1,259 | 0.68–6.27 | 0.32–50 $cm^3/g$ | – | – | Mercury uptake | When elemental sulfur was used in the impregnation process, the final level of sulfur content decreased as the impregnation temperature increased; for $H_2S$ impregnation, however, the final sulfur content increased with the impregnation temperature; sulfur-impregnated ACs were found to be quite effective in removing $Hg^0$ from gas streams |
| 150–670 | – | 0.078–0.345 $cm^3/g$ | 140 | – | $Hg^0$ adsorption capacity | For sulfur-impregnated ACs, surface area increased significantly with increasing impregnation temperatures up to a specific point and remained almost constant afterwards; however, the surface areas of the impregnated ACs were lower than those of virgin ones; the same effect was also observed for pore volume; sulfur-impregnated ACs performed significantly better than their virgin counterparts in removing $Hg^0$ from the gas stream; the $Hg^0$ adsorption capacity of the impregnated ACs markedly increased with increasing impregnation temperatures (and therefore surface area and pore volume); extended impregnation times also significantly increased the $Hg^0$ adsorption capacity of the impregnated ACs |

Below, we address the efficiency of sulfur-impregnated ACs, compared to virgin ones, for gas-phase $Hg^0$ removal. In addition, in the following subsections, we separately address the effect of influential operational parameters on Hg removal efficiency.

## 3.1 The Removal Efficiency of Sulfur-Impregnated ACs vs. Virgin ACs

In the present work, it was not possible to compare $Hg^0$ adsorption capacities of sulfur-impregnated ACs between studies, because each study we reviewed had adopted different points on the breakthrough curves for calculating $Hg^0$ adsorption capacity. Therefore, a single pooled value representing the mean $Hg^0$ adsorption capacity could not be obtained. For example, in one study, a $c/c_0$ ratio (outlet to initial $Hg^0$ concentration ratio) of 0.6 had been selected as the breakthrough point, whereas a ratio of 1 had been used in other studies. Second, as can be seen from Table 1, the impregnation settings and the operational conditions applied in different studies were so varied that inter-study comparisons of $Hg^0$ adsorption capacities would be invalid. The reason is that $Hg^0$ adsorption rates are dependent upon both impregnation settings and operational conditions.

However, all included studies indicated significantly higher $Hg^0$ adsorption capacities for sulfur-impregnated ACs than for virgin ones (Feng et al. 2006a, b, c; Granite et al. 2000; Ho et al. 2004; Hsi et al. 1998, 2001, 2002; Karatza et al. 2000; Korpiel and Vidic 1997; Krishnan et al. 1994; Kwon and Vidic 2000; Lee and Park 2003; Liu et al. 1998, 2000; Lu et al. 2011; Nabais et al. 2006; Otani et al. 1988; Skodras et al. 2005; Vidic 1996; Vidic et al. 1998; Vidic and McLaughlin 1996; Vitolo and Pini 1999; Yan et al. 2004). From the different operational conditions applied, the sulfur-impregnated ACs exhibited a range of $Hg^0$ adsorption capacities that were between 1.5 (e.g., (Hsi et al. 1998)) and 32 (e.g., (Hsi et al. 2001)) times higher than those of virgin ACs. This is primarily because $Hg^0$ captured by virgin ACs follows a physisorption mechanism (Krishnan et al. 1994), whereas that captured by sulfur-impregnated ACs occurred by a combination of physisorption of $Hg^0$ on carbon texture and chemical reactions between $Hg^0$ and impregnated sulfur, with subsequent formation of HgS (the latter being the dominant mechanism) (Steijns et al. 1976).

Therefore, capturing $Hg^0$ with virgin ACs depends primarily on AC surface area. Krishnan et al. (1994) suggested that using either a higher AC surface area or sulfur-impregnated AC effectively captured $Hg^0$. However, S impregnation significantly decreases the surface area of ACs from sulfur deposition. The greater $Hg^0$ adsorption capacities of S-impregnated ACs exceed those of virgin ACs that have higher surface areas. Hsi et al. (1998) suggested that, although sulfur impregnation of an AC decreased its surface area from 505 to 427 $m^2/g$ (a 14% decrease), its $Hg^0$ adsorption capacity nevertheless increased from 1,304 to 2,051 µg $Hg^0$/g C (>50% increase), and this increase was due to a 11% increase in the sulfur content.

## 3.2 Effect of Operational Temperature

Studies have indicated that the operational (also called reaction or bed) temperature has a pronounced effect on $Hg^0$ adsorption capacities of both virgin and sulfur-impregnated ACs (Granite et al. 2000; Karatza et al. 2000; Korpiel and Vidic 1997; Krishnan et al. 1994; Lee and Park 2003; Skodras et al. 2005; Vidic 1996; Vidic et al. 1998; Vidic and McLaughlin 1996; Vitolo and Pini 1999; Yan et al. 2004). In particular, a significant decrease in the $Hg^0$ adsorption capacity occurs as operating temperatures increase; the effect is more intense at operating temperatures higher than 100 °C (Granite et al. 2000; Krishnan et al. 1994; Skodras et al. 2005; Vidic 1996; Vidic et al. 1998; Vidic and McLaughlin 1996; Yan et al. 2004). Krishnan et al. (1994), for example, indicated that the $Hg^0$ adsorption capacities of virgin ACs halved as the temperature increased from 23 to 140° C. The authors suggested that this behavior resulted from the physisorption mechanism of $Hg^0$ capture by virgin ACs. They also suggested that altering surface properties (i.e., deactivation of surface sites that capture $Hg^0$) at higher temperatures might contribute to this effect (Krishnan et al. 1994). It is noteworthy that higher temperatures produced faster kinetics of $Hg^0$ capture; thus, Hg breakthrough is achieved in a shorter time (Vidic et al. 1998).

For sulfur-impregnated ACs, some authors have reported increased $Hg^0$ adsorption capacities with increasing operational temperatures, mainly from the chemisorptive nature of $Hg^0$ capture by sulfur-impregnated ACs (Granite et al. 2000; Korpiel and Vidic 1997; Krishnan et al. 1994; Lee and Park 2003; Vidic 1996; Vidic and McLaughlin 1996; Vitolo and Pini 1999). Others have reported either no or a negative impact (Ho et al. 2004; Karatza et al. 2000; Skodras et al. 2005; Vidic et al. 1998; Yan et al. 2004). However, these differences in results are primarily attributed to varied ranges of operational temperatures that were applied in different studies. Closer inspection of the study results revealed that up to temperatures of about 100° C, the $Hg^0$ adsorption capacities of all AC types increased as temperature increased. This behavior resulted from the chemisorptive nature of $Hg^0$ capture by sulfur-impregnated ACs and the subsequently improved kinetics of HgS formation from the increased temperature (Vidic and McLaughlin 1996). The effect of temperatures above 100° C, however, depended primarily on the type of AC applied. For commercially available sulfur-impregnated ACs, which are believed to be impregnated at temperatures between 150 and 200° C (Lee and Park 2003), any further increase in operational temperature above 100° C (higher than the melting point of sulfur, i.e., 115.2 °C) deteriorates $Hg^0$ adsorption capacity. The main reason for this is that high concentrations lead to melting and agglomeration at the weakly bonded sulfur to carbon surface boundary (i.e., the liquid being present in the form of long polymer chains (Hampel 1968)), which in turn decrease the available surface area for $Hg^0$ capture. For ACs impregnated at high temperatures (600° C, for instance), however, increasing the operational temperature above 100° C does not deteriorate the $Hg^0$ adsorption capacity. This is primarily because the bonding of sulfur to carbon is much stronger and the sulfur is more evenly distributed into the carbon matrix (Korpiel and Vidic 1997). We address the effect of impregnation temperature in-depth in Sect. 3.5.

## 3.3 Effect of Inlet $Hg^0$ Concentration

Certain findings on the effect of inlet $Hg^0$ concentration on $Hg^0$ adsorption capacity of both virgin and sulfur-impregnated ACs are controversial. A vast majority of authors have reported increased $Hg^0$ adsorption capacities at higher inlet $Hg^0$ concentrations (Korpiel and Vidic 1997; Krishnan et al. 1994; Vidic et al. 1998; Vidic and McLaughlin 1996). Such behavior is attributed to the natural driving force of concentration; Jozewicz and Gullett (1993) suggested that higher inlet $Hg^0$ concentrations would produce higher $Hg^0$ uptakes, mainly from providing a higher driving force. In addition, faster kinetics of $Hg^0$ capture by ACs can be achieved by applying higher inlet $Hg^0$ concentrations (Vidic et al. 1998). Two studies, however, have reported either no significant effect or a negative impact on $Hg^0$ adsorption capacity at higher inlet $Hg^0$ concentrations (Karatza et al. 2000; Yan et al. 2004). It was suggested that this behavior occurs primarily with commercially available sulfur-impregnated ACs, in which sulfur is primarily deposited in the macropores of ACs rather than in micropores. When this occurs, the sulfur molecules at the surface can easily become saturated with $Hg^0$ molecules at higher $Hg^0$ concentrations, making the sulfur in the bulk carbon unavailable for $Hg^0$ molecules due to their low diffusion rate into the sulfur matrix (Korpiel and Vidic 1997; Vidic et al. 1998). Nevertheless, the existing evidence appears to favor the former effect, i.e., increased $Hg^0$ adsorption capacities at higher inlet $Hg^0$ concentrations.

## 3.4 Effect of Sulfur-to-Carbon (S/C) Ratio and Sulfur Content

Although surface area and pore volume are the most influential characteristics of virgin ACs, sulfur content also significantly affects $Hg^0$ adsorption capacity of sulfur-impregnated ACs. Authors have reported that increasing the initial S/C ratio during the impregnation process significantly increases the final sulfur content as well as the sorption capacity of the impregnated ACs (Liu et al. 1998, 2000). However, this effect is strong up to a ratio of about 2/1, whereas it becomes less significant at higher ratios. For example, Liu et al. (1998) reported that increasing the initial S/C ratio from 1/4 to 2/1 increased the sulfur content from 7.17 to 10.11% (an approximately 60% rise) and the $Hg^0$ adsorption capacity from approximately 105 µg/g to about 2,050 µg/g (slightly less than a 100% increase) (Liu et al. 1998). However, a further increase of the S/C ratio to 4/1 did not exhibit this impact; the final sulfur content and the $Hg^0$ adsorption capacity rose slightly to 10.45% and about 2,250 µg/g, respectively (Liu et al. 1998). It is also noteworthy that the surface area decreased as the initial S/C ratios increased during impregnation (Liu et al. 1998).

Several authors have reported good correlations between the sulfur content of impregnated ACs and their $Hg^0$ adsorption capacities (Feng et al. 2006a, c; Hsi et al. 1998, 2002; Kwon and Vidic 2000; Lee and Park 2003; Skodras et al. 2005; Vitolo and Pini 1999). Up to a sulfur content of about 10–20%, the $Hg^0$ adsorption capacity increased with rising sulfur content. As mentioned in Sect. 3.1, this is mainly because of strong chemical bonds between Hg and S on the carbon surface, which is the major mechanism for $Hg^0$ capture by sulfur-impregnated ACs (Anton Lopez et al. 2002). In fact, sulfur atoms that exist on the carbon surface accept electrons from $Hg^0$ atoms and improve the electrode characteristics of the carbon surface (Bansal et al. 1988; Li et al. 2003). However, above a sulfur content of about 10–20%, the $Hg^0$ adsorption capacity of sulfur-impregnated ACs deteriorates significantly. This is believed to result primarily from the fact that at lower levels (i.e., up to 20%), the sulfur deposited on the carbon surface is available for binding with $Hg^0$, but at higher levels, sulfur stratification may render the sulfur unavailable for binding with $Hg^0$ and subsequent HgS formation (Vitolo and Pini 1999). This finding was confirmed by Vitolo and Pini (1999), who observed a significant decrease in the fraction of sulfur reacted with $Hg^0$ as S content increased (Vitolo and Pini 1999). Another reason might be the blockage of carbon micropores by sulfur deposition.

The effect of adding sulfur and pore volume to the carbon surface area has also been well studied (Hsi et al. 2002; Kwon and Vidic 2000; Vitolo and Pini 1999). The authors found that surface area and pore volume significantly decreased as sulfur content increased. Hsi et al. (2002) found an almost-linear association between the amount of sulfur added to the carbon texture with surface area ($R^2 = 0.73$) and pore volume ($R^2 = 0.78$). This effect is believed to result from the filling of carbon pore volume by sulfur molecules.

Although $Hg^0$ capture by sulfur-impregnated ACs occurs from chemisorption of $Hg^0$ through HgS formation, under some conditions (e.g., lower temperatures), the physisorption mechanism also becomes important (Bylina et al. 2009). This is probably because some associations exist between the surface area and $Hg^0$ adsorption capacity of sulfur-impregnated ACs (Feng et al. 2006c; Hsi et al. 2002; Skodras et al. 2005). However, the extent of this association is considerably smaller than that observed for sulfur content. Surface area becomes important only when the sulfur content among different ACs is similar.

## 3.5 *Effect of Impregnation Temperature and Time*

Impregnation temperature is one of the most critical parameters influencing carbon characteristics and, therefore, its $Hg^0$ adsorption capacity. Studies have indicated that for ACs impregnated with elemental sulfur at temperatures of up to 400–600 °C, any increase in the impregnation process decreases both the sulfur content and the loss of surface area during impregnation (Hsi et al. 2001; Korpiel and Vidic 1997; Kwon and Vidic 2000; Lee and Park 2003; Lu et al. 1998, 2011). Liu et al. (1998), for example, indicated that AC samples impregnated at 600° C had their sulfur content and surface areas in the range of 10.04–10.18% and 813.7–845.7 $m^2/g$, respectively, whereas the corresponding values for those impregnated at 250° C were 36.2–38.5% and 164.4–170.6 $m^2/g$. Kwon and Vidic (2000) suggested that this is due mainly to the effect of temperature on sulfur allotropes present on the carbon surface. At higher temperatures, sulfur exists primarily in the form of $S_2$ and $S_6$ linear chains, which are quite small and can therefore penetrate into the narrower pores of the carbon matrix; this results in few pore blockages and, in turn, little loss in surface area during the impregnation process. At lower temperatures (e.g., 200° C), sulfur primarily exists in the form of $S_7$ and $S_8$ rings, which can only enter large pores, wherein they form clusters. Formation of these clusters blocks pore entrances, which ultimately decrease the available surface area (Liu et al. 1998).

Although increasing impregnation temperatures decreases the sulfur content, the $Hg^0$ adsorption capacity of ACs impregnated at higher temperatures are significantly higher than those impregnated at lower temperatures (Feng et al. 2006a, b, c; Hsi et al. 2001; Korpiel and Vidic 1997; Kwon and Vidic 2000; Lee and Park 2003; Liu et al. 1998; Lu et al. 2011; Nabais et al. 2006). For example, Liu et al. (1998) studied the $Hg^0$ adsorption capacity of the AC impregnated at 250 °C and found that it was 550 μg/g, whereas the capacity of the AC impregnated at 600 °C was 2,200 μg/g (300% increase). This effect can be attributed to several reasons. First, in contrast to sulfur content, surface area increases with rising impregnation temperatures, which can facilitate mercury capture (Liu et al. 1998). Second, as mentioned above, sulfur allotropes present at 600 °C ($S_2$ and $S_6$ chains) are smaller and more easily penetrate into the narrow pores of the AC, while $S_8$ rings present at 200 °C tend to form clusters, which only enter large pores, blocking narrower ones. Therefore, the sulfur impregnated at 600 °C is expected to be more evenly distributed in the pore structure of ACs, while the sulfur impregnated at lower temperatures is most likely to be condensed on the external surface of the carbon (Korpiel and Vidic 1997). Third, $S_2$-to-$S_6$ chains are much more reactive than $S_8$ rings, mainly because they encompass more sulfur terminal atoms (Daza et al. 1991). Finally, thermogravimetric analyses (TGA) have indicated that when sulfur-impregnated ACs are subjected to significantly higher temperatures (as high as 400–100 °C), the percent of sulfur loss from ACs impregnated at higher temperatures is negligible. In contrast, those impregnated at lower temperatures lose a major fraction of their sulfur (as much as 90%), which is due to the stronger binding of sulfur and carbon at higher impregnation temperatures (Hsi et al. 2001; Korpiel and Vidic 1997; Kwon and Vidic 2000; Liu et al. 1998).

Several authors have suggested that increases above 600 °C in the impregnation temperature deteriorate $Hg^0$ adsorption capacity, although these ACs posses a high sulfur content (Feng et al. 2006a, b; Granite et al. 2000; Hsi et al. 2001; Nabais et al. 2006). The suggested mechanism behind this behavior is that the sulfur molecules present on the surface of ACs impregnated above 600 °C have already reacted with metals or other compounds and, therefore, are not available for $Hg^0$ capture (Feng et al. 2006a). In addition, studies have also indicated that when $H_2S$ rather than elemental sulfur is used for impregnation, the sulfur content increases with increasing impregnation temperatures (Feng et al. 2006a, b, c; Nabais et al. 2006).

Impregnation time also has a significant impact on both sulfur content and surface area, with prolonged impregnation times markedly increasing the former while dramatically deteriorating the latter (Feng et al. 2006c; Kwon and Vidic 2000;

Lu et al. 2011; Nabais et al. 2006). However, the effect of impregnation time on $Hg^0$ adsorption capacity is also temperature dependent. At lower impregnation temperatures (200 °C), extended impregnation times significantly deteriorate the $Hg^0$ adsorption capacity (Kwon and Vidic 2000); in contrast, at higher temperatures, $Hg^0$ capture increases at prolonged impregnation times (Lu et al. 2011). The reasons for this behavior are most likely similar to those observed for $Hg^0$ capture behavior of ACs impregnated at different temperatures.

# 4 Summary

This systematic review of high-quality, relevant original research articles existing in the literature was conducted to comprehensively explore the efficiency of $Hg^0$ capture from stack emissions by sulfur-impregnated vs. virgin ACs. Our systematic overview suggested that significantly higher amounts of $Hg^0$ are absorbed by sulfur-impregnated ACs than by virgin ones (1.5–32 times higher, based on the applied operational conditions). The main reason for this is because $Hg^0$ capture by virgin ACs follows a physisorption mechanism, whereas that by sulfur-impregnated ACs occurs from a combination of physisorption of $Hg^0$ on carbon texture and chemical reaction between $Hg^0$ and impregnated sulfur, with subsequent formation of HgS. Temperature increased the $Hg^0$ adsorption capacity of virgin ACs, especially when temperatures exceeded 100 °C. For sulfur-impregnated ACs, increasing the temperature up to 100 °C increased the $Hg^0$ adsorption capacity by enhancing the chemisorption of $Hg^0$ capture. A further increase in temperature enhanced the efficiency of ACs that were impregnated with S at higher temperatures (600 °C, for instance). This mainly resulted from production of stronger bonding of sulfur to carbon at higher impregnation temperatures and also from a more even distribution of sulfur in the carbon matrix.

The authors of different papers reported different results with respect to whether there is an effect of initial $Hg^0$ concentration on AC adsorption capacity. The authors of two studies could find no such effect. The predominant evidence, however, favors the view that increased $Hg^0$ adsorption capacities exist at higher inlet $Hg^0$ concentrations. Such behavior is attributed to faster kinetics of $Hg^0$ capture and an enhanced higher driving force at higher initial $Hg^0$ inlet concentrations. Results from reviewed studies also indicated that the optimum S/C ratio and sulfur content are 2/1 and 10–20%, respectively. Surface area has a less significant impact on $Hg^0$ adsorption capacity than does sulfur content. However, at equivalent sulfur content, AC surface area also becomes an important factor, in that $Hg^0$ adsorption capacity is accentuated at higher surface areas.

We conclude from having prepared this review that sulfur-impregnated ACs have significantly greater efficiencies than virgin ACs for capturing $Hg^0$ from stack emissions. Therefore, using them is more cost effective than using raw ACs; using them can also partly resolve the problem of high costs posed by applying carbon sorbents. In addition, the sulfur deposited in the ACs impregnated at higher temperatures is more evenly distributed in the carbon micropores and binds more strongly to the carbon matrix. Hence, sulfur-impregnated ACs can retain higher $Hg^0$ adsorption capacities under actual stack conditions, if the temperature is at least 140 °C. Finally, since the major mechanism for $Hg^0$ removal by sulfur-impregnated ACs is through the chemical reaction between $Hg^0$ and S, and subsequent formation via strong bonds of HgS, the $Hg^0$ adsorbed on ACs is quite stable and is not easily released when discharged as waste to the environment.

**Acknowledgements** This research has been supported by Tehran University of Medical Sciences and Health Services grant (project no. 90-04-46-16808). Hereby, the cooperation of the University and Institute for Environmental Research (IER) is highly appreciated.

# References

Anton Lopez AM, Tascon MDJ, Martinez-Tarazona MR (2002) Retention of mercury in activated carbons in coal combustion and gasification flue gases. Fuel Process Technol 77–78:353–358
Bansal CR, Donnet BJ, Stoeckli F (1988) Active carbon. Marcel Dekker, New York
Berlin M (1979) Mercury. In: Friberg L, Nordberg GF, Vouk VB (eds) Handbook on toxicology of metals. Elsevier, Amsterdam
Brenneman KA, James RA, Gross EA, Dorman DC (2000) Olfactory neuron loss in adult male CD rats following subchronic inhalation exposure to hydrogen sulfide. Toxicol Pathol 28:326–333
Bylina IV, Tong S, Jia CQ (2009) Thermal analysis of sulphur impregnated activated carbons with mercury adsorbed from the vapour phase. J Therm Anal Calorim 96:91–98
Cal MP, Strickler BW, Lizzio AA (2000) High temperature hydrogen sulfide adsorption on activated carbon I. Effects of gas composition and metal addition. Carbon 38:1757–1765
Clakson T (1993) Health effects associated with mercury contamination. In: Proceedings of the 1993 international municipal waste combustion conference, Williamsburg, VA, pp 591–597
Daza L, Mendioroz S, Pajares JA (1991) Mercury adsorption by sulfurized fibrous silicates. Clays Clay Miner 39:14–21
Dorman DC, Moulin FJM, McManus BE, Mahle KC, James RA, Struve MF et al (2002) Cytochrome oxidase inhibition induced by acute hydrogen sulfide inhalation: correlation with tissue sulfide concentrations in the rat brain, liver, lung, and nasal epithelium. Toxicol Sci 65:18–25
Feng W, Borguet E, Vidic RD (2006a) Sulfurization of a carbon surface for vapor phase mercury removal—II: Sulfur forms and mercury uptake. Carbon 44:2998–3004
Feng W, Borguet E, Vidic RD (2006b) Sulfurization of carbon surface for vapor phase mercury removal—I: Effect of temperature and sulfurization protocol. Carbon 44:2990–2997
Feng W, Kwon S, Feng X, Borguet E, Vidic RD (2006c) Sulfur impregnation on activated carbon fibers through $H_2S$ oxidation for vapor phase mercury removal. J Environ Eng 132:292–300
Galbreath KC, Zygarlicke C (1996) Mercury speciation in coal combustion and gasification flue gases. Environ Sci Technol 30:2421–2426
Granite EJ, Pennline HW, Hargis RA (2000) Novel sorbents for mercury removal from flue gas. Ind Eng Chem Res 39:1020–1029
Green BN, Johnson CD, Adams A (2001) Writing narrative literature reviews for peer-reviewed journals: secrets of the trade. J Sports Chiropr Rehabil 15:5–19
Hampel CA (ed) (1968) The encyclopedia of the chemical elements. Reinhold Book Corporation, New York
Ho TC, Kobayashi N, Lee Y, Lin J, Hopper JR (2004) Experimental and kinetic study of mercury adsorption on various activated carbons in a fixed-bed adsorber. Environ Eng Sci 21:21–27

Hsi HC, Chen SG, Rostam-Abadi M, Rood MJ, Richardson CF, Carey TR, Chang R (1998) Preparation and evaluation of coal-derived activated carbons for removal of mercury vapor from simulated coal combustion flue gases. Energy Fuel 12:1061–1070

Hsi HC, Rood MJ, Rostam-Abadi M, Chen SG, Chang R (2001) Effects of sulfur impregnation temperature on the properties and mercury adsorption capacities of activated carbon fibers (ACFs). Environ Sci Technol 35:2785–2791

Hsi HC, Rood MJ, Rostam-Abadi M, Chen SG, Chang R (2002) Mercury adsorption properties of sulfur-impregnated adsorbents. J Environ Eng 128:1080–1089

Johnson J (1997) Controversial EPA mercury study endorsed by science panel. Environ Sci Technol 31:218A–219A

Jozewicz W, Gullett BK (1993) The 1993 international conference on managing hazardous air pollution. Electric Power Research Institute, Palo Alto, CA, pp VII-85–VII-99

Karatza D, Lancia A, Musmarra D, Zucchini C (2000) Study of mercury absorption and desorption on sulfur impregnated carbon. Exp Therm Fluid Sci 21:150–155

Khan KS, Kunz R, Kleijen J, Antes G (2003) Systematic reviews to support evidence-based medicine: how to write and apply findings of healthcare research. Royal Society of Medicine Press, London

Korpiel JA, Vidic RD (1997) Effect of sulfur impregnation method on activated carbon uptake of gas-phase mercury. Environ Sci Technol 31:2319–2325

Krishnan SV, Gullett BK, Jozewicz W (1994) Sorption of elemental mercury by activated carbons. Environ Sci Technol 28:1506–1512

Kwon S, Vidic RD (2000) Evaluation of two sulfur impregnation methods on activated carbon and bentonite for the production of elemental mercury sorbents. Environ Eng Sci 17:303–313

Lee SH, Park YO (2003) Gas-phase mercury removal by carbon-based sorbents. Fuel Process Technol 84:197–206

Li HY, Lee WC, Gullett KB (2003) Importance of activated carbon's oxygen surface functional groups on elemental mercury adsorption. Fuel 82:451–457

Liu W, Vidic RD, Brown TD (1998) Optimization of sulfur impregnation protocol for fixed bed application of activated carbon-based sorbents for gas-phase mercury removal. Environ Sci Technol 32:531–538

Liu W, Vidic RD, Brown TD (2000) Optimization of high temperature sulfur impregnation on activated carbon for permanent sequestration of elemental mercury vapors. Environ Sci Technol 34:483–488

Lu C, Liu Q, Gao W, Yang J (2011) Microstructure and surface morphology of sulfur-loaded activated carbons for mercury removal in coal-fired flue gas. Fresen Environ Bull 20:135–139

Nabais JV, Carrott PJM, Carrott MMLR, Belchior M, Boavida D, Diall T, Gulyurtlu I (2006) Mercury removal from aqueous solution and flue gas by adsorption on activated carbon fibres. Appl Surf Sci 252:6046–6052

Otani Y, Emi H, Kanaoka C, Uchijima I, Nishino H (1988) Removal of mercury vapor from air with sulfur-impregnated adsorbents. Environ Sci Technol 22:708–711

Pacyna JM, Munch J (1991) Anthropogenic mercury emission in Europe. Water Air Soil Pollut 5:51–61

Pavlish JH, Sondreal EA, Mann MD, Olson ES, Galbreath KC, Laudal DL, Benson SA (2003) Status review of mercury control options for coal-fired power plants. Fuel Process Technol 82:89–165

Schroedor WH, Yarwood G, Niki H (1991) Involving mercury species in the atmosphere. Water Air Soil Pollut 56:653–666

Schuster E (1991) The behavior of Hg in soil with special emphasis on complexation and adsorption process. Water Air Soil Pollut 56:667–680

Sinha RK, Walker PL (1972) Removal of mercury by sulfurized carbons. Carbon 10:754–756

Skodras G, Diamantopoulou I, Natas P, Palladas A, Sakellaropoulos GP (2005) Postcombustion measures for cleaner solid fuels combustion: activated carbons for toxic pollutants removal from flue gases. Energy Fuel 19:2317–2327

Steijns M, Peppelenbos A, Mars P (1976) Mercury chemisorption by sulfur adsorbed in porous materials. J Colloid Interface Sci 57:181–186

Vidic RD (1996) Control of mercury emissions in flue gases by activated carbon adsorption. ACS Division of Fuel Chemistry, Preprints, vol 41, pp 437–439

Vidic RD, McLaughlin JB (1996) Uptake of elemental mercury vapors by activated carbons. J Air Waste Manag Assoc 46:241–250

Vidic RD, Chang MT, Thurnau RC (1998) Kinetics of vapor-phase mercury uptake by virgin and sulfur-impregnated activated carbons. J Air Waste Manag Assoc 48:247–255

Vitolo S, Pini R (1999) Deposition of sulfur from $H_2S$ on porous adsorbents and effect on their mercury adsorption capacity. Geothermics 28:341–354

Volland C (1991) 84th Annual meeting and exhibition. AWMA, Vancouver, BC, Paper 91-35.91

Weast RC (ed) (1983) CRC handbook of chemistry and physics. CRC Press, Boca Raton, FL

Yan R, Liang DT, Tsen L, Wong YP, Lee YL (2004) Bench-scale experimental evaluation of carbon performance on mercury vapour adsorption. Fuel 83:2401–2409

Young BC, Miller SJ, Laudal DL (1994) The 1994 Pittsburgh coal conference, Pittsburgh, PA

# Setting Water Quality Criteria in China: Approaches for Developing Species Sensitivity Distributions for Metals and Metalloids

Yuedan Liu, Fengchang Wu, Yunsong Mu, Chenglian Feng, Yixiang Fang, Lulu Chen, and John P. Giesy

## Contents

1 Introduction .................................................................................................................. 36
2 Data Selection and Analysis ........................................................................................ 38
   2.1 Data Collection ..................................................................................................... 38
   2.2 Methods Used to Construct SSDs ........................................................................ 40
   2.3 Risk Assessment Procedure ................................................................................. 43
3 SSD Construction and Model Comparison ................................................................. 43
   3.1 Hazardous Concentration (HC5) ......................................................................... 43
   3.2 Comparison of Approaches .................................................................................. 44
   3.3 Species Sensitivity ............................................................................................... 46
4 Risk Assessments ........................................................................................................ 49
   4.1 Measured Exposure Concentrations .................................................................... 49
   4.2 Correlation Analysis ............................................................................................ 49
   4.3 Hazard Quotients ................................................................................................. 50

Y. Liu
State Key Laboratory of Environmental Criteria and Risk Assessment,
Chinese Research Academy of Environmental Sciences, Beijing 100012, China

Environmental Simulation and Pollution Control Research Center,
South China Institute of Environmental Sciences, Guangzhou 510065, China

F. Wu (✉) • Y. Mu • C. Feng • Y. Fang • L. Chen
State Key Laboratory of Environmental Criteria and Risk Assessment,
Chinese Research Academy of Environmental Sciences, Beijing 100012, China
e-mail: wufengchang@vip.skleg.cn

J.P. Giesy
Department of Veterinary Biomedical Sciences and Toxicology Centre,
University of Saskatchewan, Saskatoon, SK, Canada

Zoology Department and Center for Integrative Toxicology, Michigan State University,
East Lansing, MI 48824, USA

| 5 | Discussion | | 51 |
|---|---|---|---|
| | 5.1 | Evaluation of Approaches | 51 |
| | 5.2 | Selection of Approaches | 52 |
| | 5.3 | Proportion of Species to Be Protected | 53 |
| | 5.4 | HQ for Risk Assessment | 53 |
| 6 | Summary | | 53 |
| References | | | 54 |

# 1 Introduction

Water quality criteria (WQCs) refer to the maximum acceptable concentrations of specific chemicals or magnitudes of parameters in water that protect aquatic life and human health under certain conditions (USEPA 1976). When deriving WQC for use in regional ecosystems, sociopolitical and economic factors need to be considered (Meng and Wu 2010). The WQC concept is often used for making policy, managing the environment, assessing water quality, controlling pollution, restoring ecosystems, and managing environmental crises (Wu et al. 2010). Some countries and organizations have created WQC guidelines that describe what is suitable for the specific conditions prevalent in that country or region. Since the 1960s, the United States has undertaken a series of long-term studies to develop national WQC for specific water pollutants that threaten aquatic organisms and human health (USEPA 1968, 1976, 1986, 1999, 2002, 2004, 2009). In the past few decades, Australia, Canada, the European Union, the Netherlands, and the World Health Organization have, respectively, developed their own WQCs to protect national or regional water environments (CCME 1999; ANZECC and ARMCANZ 2000; ECB 2003; WHO 2006; RIVM 2007).

As a country with rich aquatic species and vast freshwater regions, China also plays an important role in protecting its share of the world aquatic ecosystems. However, the water environment of China is suffering from contamination with metals and metalloids that has and is being released from human activities; much of this contamination is a consequence of China's rapid economic development and the expansion of its human population. Like other countries, China manages water quality by establishing or adopting water quality standards. However, the WQC standards for other countries may not be wholly appropriate for conditions in China. WQCs that are specific to certain geographic regions, other than China, and the species composition therein may not be appropriate for managing the environment in China. Thus, it is urgent to establish guidelines or WQCs that are based on the characteristics, composition, and distribution of aquatic species endemic to China.

Information on the toxicity of chemicals to aquatic organisms that is applied in ecological risk assessments usually comes from tests with single species. However, the entity to be protected is not limited to individuals but rather extends to populations, communities, and ecosystems. Species sensitivity distributions (SSDs) are useful for extrapolating between the macro-scale (such as communities) and the microscale (such as individuals) in an integrative risk assessment across temporal and spatial scales (Newman et al. 2000). As an efficient tool for ecological risk assessment, SSDs have received considerable attention since the 1980s (Stephan et al. 1985; Aldenberg and Slob 1993). SSDs are used to investigate relationships among sensitivities of species to environmental stressors, such as metals and organic chemicals. The primary purpose of establishing SSDs is to determine the tolerated concentration of a substance for protecting individuals of a defined proportion of a species found in an assemblage (usually 95%), and this tolerated concentration may be hazardous to 5% of total species (HC5) (van Straalen and Denneman 1989). For this purpose, SSDs are visualized as a plot of a cumulative distribution function against the logarithm of the concentrations of toxicity data (Solomon et al. 2000). Also, SSDs offer greater statistical confidence in calculating a predicted no effect concentration (PNEC) for use in risk assessments than does the commonly used quotient approaches (Grist et al. 2002; Wheeler et al. 2002; Wang et al. 2008). The latter approaches are usually calculated by applying a safety factor to the statistical summary of a single toxicity test such as no observed effect concentration (NOEC) or a 50%-effect concentration ($EC_{50}$) (van Dam et al. 2012).

When constructing SSDs, there are no standard approaches to achieve fits to all toxicity data. However, several approaches have been applied to develop SSDs and to estimate HC5 values, which include Burr Type III (Shao 2000), Gompertz (Newman et al. 2000), log-logistic, lognormal (Pennington 2003), and Weibull (van Straalen 2002). A recent study reported and compared the array of statistical distributions used to analyze air contaminant data (Marchant et al. 2013). The common characteristic of these approaches is the assumption that species sensitivities follow certain specific statistical distributions. However, this assumption is often violated due to statistical limitations resulting from deviations between theoretical and empirical data (Forbes and Forbes 1993; Calow 1996; Power and McCarty 1997; Grist et al. 2002). In practice, a majority of the data selected usually do not meet all assumptions. For instance, the most commonly used lognormal distribution failed to fit the toxicity data on a number of occasions (Newman et al. 2000). To resolve this limitation in deriving HC5 values for contaminants, without making any assumptions about the underlying distributions, use of a more robust nonparametric method, known as bootstrap, has been suggested. Bootstrap resampling methods were first used to estimate HC5 values of pesticide by constructing SSDs (Jagoe and Newman 1997; Newman et al. 2000). The bootstrap regression was further developed by combining a nonparametric bootstrap with a parametric log-logistic model to solve the difficulty of the limited toxicity data available (Grist et al. 2002). Based on the standard bootstrap, we applied artificial interpolations to avoid repetitive values in each resample and to expand the data beyond the limited original datasets (Wang et al. 2008).

Metals and metalloids (Power and McCarty 1997; Duffus 2002; Batley 2012; Chapman 2012) are widely distributed in the environment and can adversely affect the diversity and the evolution of aquatic organisms (Shaw and Grushkin 1957; Campbel and Stokes 1985; Mance 1987). For instance, cadmium is a typical metal pollutant that has been associated with many epidemiological diseases such as the *itai-itai* disease in Japan (Nogawa and Kido 1993). Although zinc is an essential element for many metabolic functions of most organisms, it is toxic to aquatic life when concentrations exceed the threshold for effects (Van Sprang et al. 2009;

Tsushima et al. 2010). The first WQC guideline for metals was developed by the USEPA in 1976; WQCs were developed for 12 metals and metalloids. Thus far, WQCs of 167 typical water pollutants have been established and these pollutants have been divided into priority and non-priority toxic classes (USEPA 1976, 1986, 1999, 2002, 2004, 2009). However, only 16 WQC values have been promulgated for protecting aquatic organisms, which include 11 for priority toxic metals and metalloids and 4 for non-priority metals (USEPA 1985; Meng and Wu 2010). SSDs have been applied in ecological risk assessments for freshwater environments, predominantly for single metals or organic pesticides. However, the reported works on SSDs have primarily focused on single metals or organic molecules (Solomon et al. 1996; Giesy et al. 1999; Campbell et al. 2000; TenBrook et al. 2010; Vardy et al. 2011). These works have not included many systematic and comparative studies on SSDs or WQC values established for multiple metals and metalloids in aquatic environments.

One goal in this study is to compare different approaches for deriving WQCs through SSDs that are based on toxicity data of representative aquatic species in China. First, we employed parametric and nonparametric approaches to develop SSDs through fitting chronic toxicity values. We evaluated sensitivities of species exposed to various chemicals before selecting indicator species for chemical biomonitoring in the water environments. We further compared the approaches by using several statistical indicators to evaluate the applicability of different approaches. Criteria for model selection were further addressed by evaluating other data parameters, including species amounts and composition, species sensitivity, and geographic structure of aquatic habitats. Differences between the WQC values we derived to meet salient needs in China were then compared to those promulgated by selected other countries. Another study goal is to determine the risk of eight metals and metalloids to Chinese aquatic species by using Tai Lake (Ch: *Taihu*) as a study area. We performed the risk assessment of the metals and metalloids to aquatic species in Tai Lake by utilizing the measured exposure concentration (MEC) and WQC values derived from SSDs created by using different approaches.

## 2 Data Selection and Analysis

### 2.1 *Data Collection*

#### 2.1.1 Toxicity Data

Chronic toxicity data for aquatic species were used for constructing SSDs. The toxicity data from the literature that was used for the species and chemicals are shown in Table 1. All data were collected from the ECOTOX database of the USEPA (http://www.epa.gov/ecotox) and the database of the China National Knowledge Infrastructure (CNKI, http://www.cnki.net/). Accuracy, reliability, and relevance of

**Table 1** Statistical summary of toxic effects of metals and metalloids on freshwater species

| Metals and metalloids | Number of species | Exposure time (days) | Log transformation of toxicity and standard deviation (SD) (µg/L) | | |
|---|---|---|---|---|---|
| | | | Geometric mean | SD | $p$-value for normality test |
| As | 17 | 8~24 | 2.46 | 0.64 | 0.652 |
| Cd | 22 | 4~36 | 0.31 | 0.15 | 0.757 |
| Cr | 27 | 4~36 | 1.64 | 0.65 | 0.366 |
| Cu | 14 | 6~24 | 2.65 | 0.67 | 0.841 |
| Hg | 26 | 4~24 | 0.59 | 0.24 | 0.724 |
| Ni | 29 | 4~18 | 2.65 | 0.42 | 0.452 |
| Pb | 28 | 4~24 | 1.64 | 0.64 | 0.566 |
| Zn | 49 | 4~36 | 2.86 | 0.37 | 0.578 |

the literature data were evaluated by using standard methods (Klimisch et al. 1997; ECB 2003). The selected metalloid was arsenic (As), and the metals included cadmium (Cd), chromium (Cr), copper (Cu), mercury (Hg), nickel (Ni), lead (Pb), and zinc (Zn) (Table 1). The species selected for developing the WQC were designed to represent examples that were widely distributed in aquatic ecosystems of China. The list included both native species and those originally imported from other countries but have now become widespread in China. The toxicity endpoints selected for deriving WQC were growth and reproductive effects. Toxicological tests of the literature data were performed according to standard operational procedures. Duration of chronic toxicity data ranged from 4 to 36 days. Toxicity threshold values were calculated and reported either as the no observed concentration (NOEC) or the lowest observed effect concentration (LOEC). Geometric means were calculated for values having multiple reports with the same exposure time (Stephan et al. 1985). When several eligible chronic toxicity data for the same species are available, the NOEC value having the longest duration of exposure was selected for use. When the NOEC value was not available, the geometric mean of the LOEC was selected. When only the LOEC value was available, we regarded half of its value as the NOEC (Balk et al. 1995).

### 2.1.2 Measured Exposure Concentrations

Surface waters were collected from 40 sites in Tai Lake during September 2010 (Fig. 1). The sampling sites were recorded by using a global positioning system. The concentrations in water of seven metals and one metalloid (viz., As, Cr, Cd, Cu, Hg, Ni, Pb, and Zn) were measured by inductively coupled plasma-optical emission spectrometry (ICP-MS, Agilent, 7500 CX, USA) and atomic fluorescence spectrophotometer (AFS, AF-610A, China). MECs were calculated and used to assess risks of the metals to aquatic species living in Tai Lake.

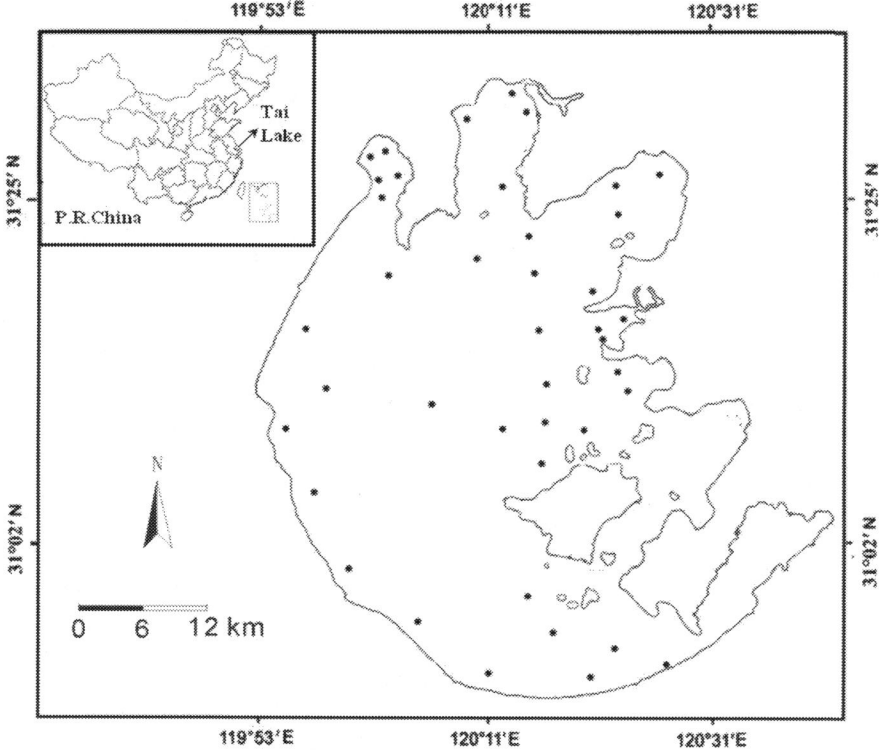

**Fig. 1** Location of 40 sampling sites in the Tai Lake (*dark points* indicate sampling sites)

## 2.2 Methods Used to Construct SSDs

### 2.2.1 Parametric Approaches

After log transformation of effect concentrations (Stephan et al. 1985; Aldenberg and Jaworska 2000; van Straalen 2002), the Shapiro–Wilk test was performed on the SPSS Version 17 software to check the normality of toxicity and their applicability to four parametric approaches, including Gompertz (Newman et al. 2000), log-logistic (Aldenberg and Slob 1993; Pennington 2003), lognormal (Wagner and Løkke 1991; Wheeler et al. 2002), and sigmoid (Wu et al. 2011). These approaches were generally applicable for fitting species sensitivity data for species toxicity datasets of metals and metalloids. The toxicity data were fitted to the four statistical distributions, and the HC5 values were generated from the curves where the cumulative probability was equal to 0.05 (Posthuma et al. 2002).

### 2.2.2 Bootstrap

The bootstrap method is a nonparametric technique for simulating any statistical distribution through resampling of observed datasets, without assuming an underlying distribution or specific curve-fitting parameters in the model (Efron and Tibshirani 1993; Varian 2005). For example, suppose that an empirical toxicity sample $t = [x_1, \ldots, x_n]$ was first randomly or independently collected from a given population. A sample of size $n$, $t_1^* = [x_{11}^*, \ldots, x_{1n}^*]$, was further drawn from the members of $t$ with replacement. Each observation $x_i$ would be sampled with an equal replacement probability of $1/n$. The sampling process was iterated $B$ times, and the $B$th bootstrap sample was marked as $t_b^* = [x_{b1}^*, \ldots, x_{bn}^*]$. The number of iterations taken in this study was set to 5,000 according to the previous report (Wang et al. 2008). The members of each bootstrap sample were sorted in ascending order. The cumulative probabilities of sorted toxicity data were calculated to derive SSDs.

The bootstrap is limited to the original observations, although it does not require any distribution for the data. It is not suitable for examining the statistical distribution of the largest or the smallest observations, since the bootstrap method never generates an observation either larger or smaller than the maximum or the minimum observation (Efron and Tibshirani 1993). The bootstrap works with discrete data to derive an HC5 value for a given dataset, although the dataset must contain at least $100/5 = 20$ members (Grist et al. 2002). Alternatively, in practice, it is difficult to collect adequate sample sizes for most cases, which restricts application of bootstrap methods. In this study, to avoid picking repetitive numbers in each resample caused by the process of the basic bootstrap, a modified bootstrap approach was developed by inserting interval values between consecutive ascending toxicity data. The modified bootstrap was applied to generate the HC5 values and was simply called *bootstrap* in this study. The detailed processing procedure was performed according to previously described methods (Wang et al. 2008).

### 2.2.3 Bootstrap Regression

The bootstrap regression was developed by combining the modified bootstrap with a log-logistic regression to improve fitting of datasets. Here we chose the log-logistic regression to combine with the bootstrap, since the conventional SSD approach yielded a good fit to the data having a log-logistic curve fitted through nonlinear regression (Grist et al. 2002) (Table 2). Bootstrapping was computed by using the modified procedure described above.

The computational processes for the parametric and nonparametric approaches were performed by the use of R programming language (Version 2.14.0, R Development Core Team 2011). Three indicators, root mean square errors (RMSE), error sum of squares (SSE), and coefficients of determination ($r^2$), were derived from the four parametric approaches, while two indicators, RMSE and SSE, were obtained from the nonparametric approaches. These indicators were used to compare outputs and check the adequacy of the candidate approaches. The model with

**Table 2** Comparison of the 5% hazardous concentration threshold (HC5) value to protect 95% of species, calculated by different approaches

| | As | Cd | Cr | Cu | Hg | Ni | Pb | Zn |
|---|---|---|---|---|---|---|---|---|
| Sample size | 17 | 22 | 27 | 14 | 26 | 29 | 28 | 49 |
| Gompertz $F(x)=ae^{-e^{(x-x_0)(-b)}}$ | | | | | | | | |
| $a$ | 1.09 | 1.15 | 1.16 | 1.67 | 0.94 | 1.05 | 0.99 | 1.02 |
| $x_0$ | 3.08 | 0.39 | 1.92 | 2.70 | 0.44 | 2.61 | 1.17 | 3.06 |
| $b$ | 1.08 | 0.99 | 0.95 | 1.54 | 0.59 | 0.92 | 0.73 | 0.91 |
| HC5 | 72.76 | 0.18 | 6.70 | 5.86 | 0.58 | 38.65 | 2.36 | 114.21 |
| $r^2$ | 0.97 | 0.98 | 0.99 | 0.94 | 0.99 | 0.99 | 0.99 | 0.99 |
| RMSE | 0.05 | 0.04 | 0.04 | 0.08 | 0.03 | 0.02 | 0.03 | 0.03 |
| SSE | 0.04 | 0.09 | 0.03 | 0.11 | 0.03 | 0.01 | 0.02 | 0.03 |
| Log-logistic $F(x)=1/[1+e^{(a-x)/b}]$ | | | | | | | | |
| $a$ | 3.54 | 5.46 | 2.15 | 2.54 | 1.38 | 3.68 | 1.79 | 3.87 |
| $b$ | 0.62 | 0.63 | 0.51 | 0.60 | 0.56 | 0.72 | 0.49 | 0.63 |
| HC5 | 50.83 | 0.23 | 4.42 | 5.89 | 0.53 | 36.03 | 2.23 | 111.9 |
| $r^2$ | 0.95 | 0.99 | 0.99 | 0.95 | 0.97 | 0.99 | 0.99 | 0.98 |
| RMSE | 0.07 | 0.03 | 0.03 | 0.07 | 0.05 | 0.02 | 0.04 | 0.04 |
| SSE | 0.06 | 0.06 | 0.02 | 0.09 | 0.07 | 0.01 | 0.03 | 0.07 |
| Lognormal (probit scale) $F(x)=ax+b$ | | | | | | | | |
| $a$ | 0.41 | 1.73 | 0.38 | 0.35 | 0.58 | 0.22 | 0.22 | 0.39 |
| $b$ | −2.42 | −0.27 | −1.98 | −1.94 | −1.48 | −1.96 | −1.75 | −2.43 |
| HC5 | 80.75 | 0.16 | 7.53 | 6.85 | 0.52 | 28.76 | 2.94 | 103.75 |
| $r^2$ | 0.95 | 0.94 | 0.87 | 0.92 | 0.81 | 0.92 | 0.94 | 0.95 |
| RMSE | 0.08 | 0.05 | 0.04 | 0.07 | 0.06 | 0.06 | 0.08 | 0.05 |
| SSE | 0.09 | 0.08 | 0.07 | 0.11 | 0.09 | 0.05 | 0.06 | 0.07 |
| Sigmoid $F(x)=1/\left[1+e^{(x-x_0)/(-a)}\right]^b$ | | | | | | | | |
| $a$ | 0.64 | 0.22 | 0.51 | 0.03 | 0.70 | 0.64 | 0.71 | 0.30 |
| $b$ | 0.32 | 0.19 | 0.41 | 0.03 | 0.19 | 0.12 | 0.27 | 0.29 |
| $x_0$ | 5.36 | 1.86 | 5.07 | 3.23 | 5.57 | 10.22 | 9.45 | 4.26 |
| HC5 | 63.53 | 0.18 | 5.75 | 4.34 | 0.47 | 26.74 | 1.96 | 92.53 |
| $r^2$ | 0.96 | 0.96 | 0.99 | 0.99 | 0.99 | 0.93 | 0.99 | 0.99 |
| RMSE | 0.07 | 0.02 | 0.03 | 0.04 | 0.04 | 0.08 | 0.03 | 0.04 |
| SSE | 0.06 | 0.02 | 0.01 | 0.02 | 0.03 | 0.16 | 0.02 | 0.05 |
| Bootstrap | | | | | | | | |
| HC5 | 92.63 | 0.23 | 8.69 | 7.36 | 0.62 | 36.32 | 3.05 | 135.23 |
| RMSE | 0.006 | 0.002 | 0.005 | 0.006 | 0.005 | 0.002 | 0.007 | 0.007 |
| SSE | 0.01 | 0.004 | 0.004 | 0.002 | 0.01 | 0.03 | 0.001 | 0.01 |
| Bootstrap regression | | | | | | | | |
| HC5 | 116.39 | 0.28 | 11.53 | 10.63 | 0.81 | 58.32 | 3.54 | 165.30 |
| RMSE | 0.002 | 0.002 | 0.003 | 0.004 | 0.001 | 0.002 | 0.003 | 0.005 |
| SSE | 0.004 | 0.004 | 0.006 | 0.008 | 0.004 | 0.004 | 0.005 | 0.006 |

The values are expressed as μg/L. Model parameters for parametric estimations are also presented

*SSE* error sum of squares, *RMSE* root mean square error

the least RMSE and SSE and greatest $r^2$ was deemed to produce the most appropriate SSDs and HC5 values. The 95% confidence intervals of HC5 values were further generated with different methods (Aldenberg and Jaworska 2000; Grist et al. 2002).

## 2.3 Risk Assessment Procedure

The hazard quotient (HQ) approach was used to screen and characterize risks posed by metals and metalloids to aquatic species in Tai Lake. Utilizing this method is consistent with the guidelines of the technical guidance document on risk assessment of the European Union (EU 1996), wherein the MEC and hazard concentration were used to obtain a PNEC. The MECs of metals and metalloids in the water samples were then compared with PNEC values to calculate the HQ as

$$HQ = MEC / PNEC. \qquad (1)$$

The PNEC is estimated by dividing the HC5 values by an uncertainty factor (UF),

$$PNEC = HC5 / UF. \qquad (2)$$

The UF value was set as 1 in this study, since the collected toxicity data were adequate to cover most of trophic levels of aquatic species (EU 1996). If the $HQ \geq 1$, a threshold for some degree of effect has been exceeded; values of $0.1 \leq HQ < 1$ indicate that a medium risk is probable; and $0.01 \leq HQ < 0.1$ indicate a low risk (Sanchez-Bayo et al. 2002).

## 3 SSD Construction and Model Comparison

### 3.1 Hazardous Concentration (HC5)

Profiles of estimated HC5 values and variables involved in the six approaches are shown in Table 2. The results of HC5 derived via six approaches were of the same order of magnitude. However, HC5 values obtained by the use of bootstrap or bootstrap regression were generally greater than those derived by using the parametric approach (Tukey's test in one-way ANOVA, $F = 525$, $DF = 5, 42$, $P < 0.001$). For instance, the HC5 value for Zn was 165.30 µg/L when derived by the use of the bootstrap regression, whereas this value was 92.53 µg/L when fitted to the log-logistic distribution. Based on overall estimates derived by the use of the various approaches, the order of decreasing toxicity of the eight elements tested was Zn < As < Ni < Cr < Cu < Pb < Hg < Cd (Fig. 2). These results are consistent with the toxicity study results on specific species with metals and metalloids (USEPA 1996).

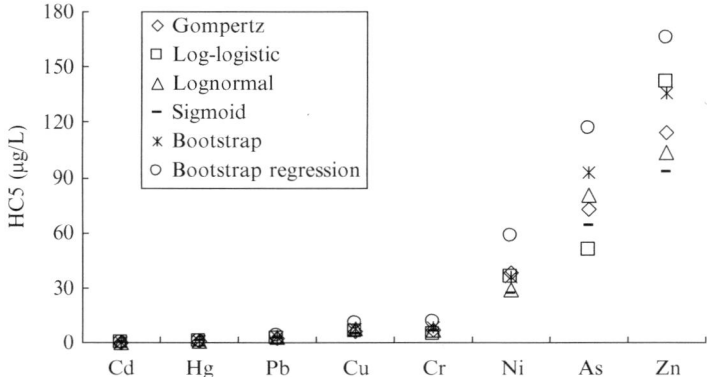

**Fig. 2** Comparison of HC5 values for metals and metalloids calculated by parametric and nonparametric approaches

HC5 values, derived from our review of available toxicity data, were compared with those published by the USEPA (1985, 1999). As an example of this comparison, HC5 values for Cd for China were in the range of 0.175–0.278 µg/L, while the USEPA determined this value to be 0.25 µg/L. Similar comparative results were observed for the five other metals evaluated (e.g., Cr, Hg, Ni, Pb, and Zn). In contrast, the HC5 values for As and Cu, published by the USEPA, were out of the range of those derived by using different approaches in this study. For example, the maximum HC5 value derived for As for China was 116.39 µg/L, which was less than the value published by the USEPA. Notwithstanding, the difference between HC5 values derived in this study for China and those published by the USEPA was reasonable, probably for two reasons. First, the freshwater ecosystems for the two countries are located on different continents. For instance, the Great Lakes in the United States are quite different from freshwater aquatic systems present in China, featuring different geographical conditions and different populations of aquatic life (Rausina et al. 2002). Second, we used different analytical approaches than did the EPA in deriving HC5 values. Specifically, USEPA generally used derivation methods that depended on the four most sensitive genera (Meng and Wu 2010), whereas we derived values by analyzing the relationship between toxicity values of tested species and their corresponding cumulative probabilities. In addition, differences in target populations and their relative contributions to the aquatic ecological system also have been responsible for differences, as well (Wu et al. 2012).

## 3.2 Comparison of Approaches

SSDs derived by the use of the bootstrap, bootstrap regression, or conventional approaches were compared (Fig. 3a–h). In general, results obtained by using bootstrap or bootstrap regression followed the empirical data points exactly, whereas

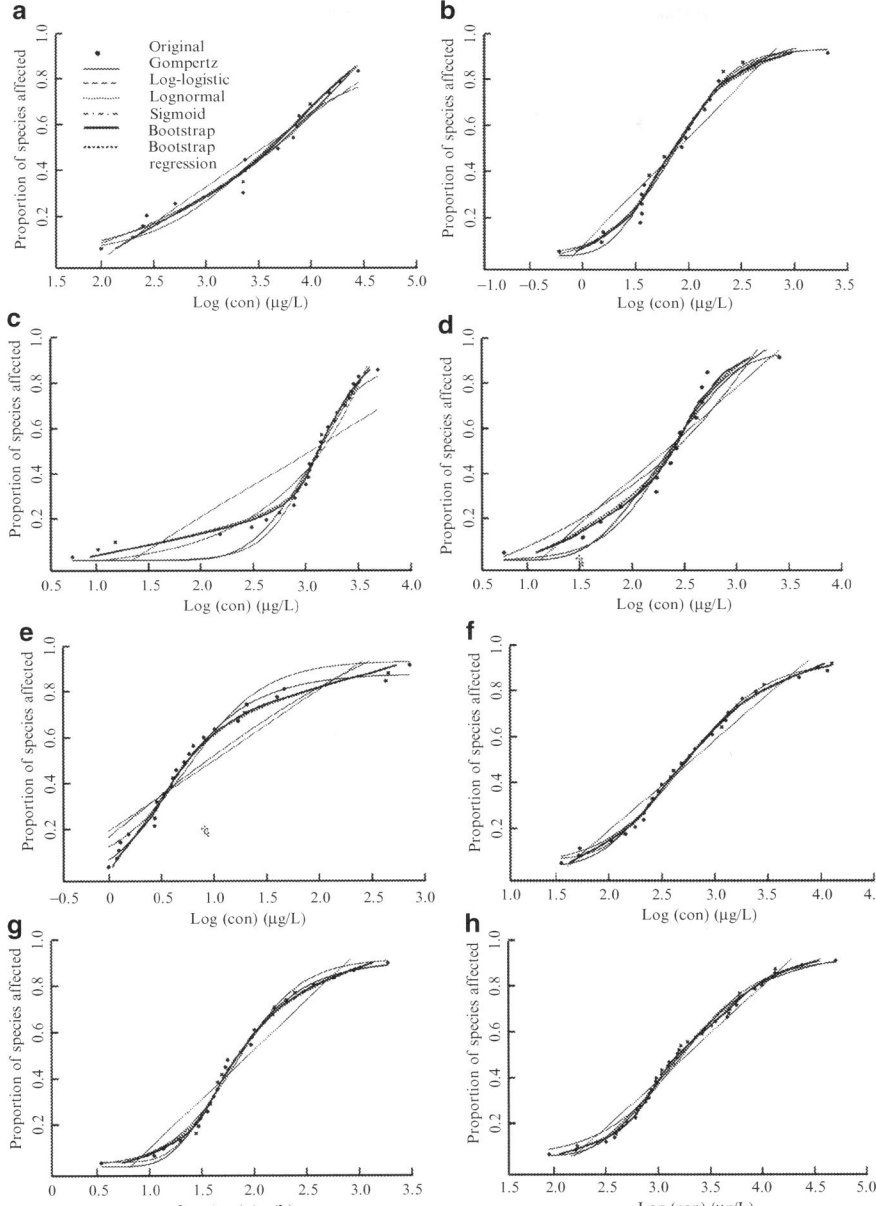

**Fig. 3** Illustration of the SSDs derived by applying different approaches. (**a**) As ($n=17$); (**b**) Cd ($n=22$); (**c**) Cr ($n=27$); (**d**) Cu ($n=14$); (**e**) Hg ($n=26$); (**f**) Ni ($n=29$); (**g**) Pb ($n=28$); (**h**) Zn ($n=49$). Legends of (**b–h**) are referred in (**a**)

some curves derived by using other methods deviated from these points. For example, the curve fitted by using the sigmoid distribution (Fig. 3c) significantly deviated from the original data. In addition, the lower tail failed to exactly follow the raw data, although the data for Cr were generally well fitted by parametric approaches such as the log-logistic distribution (Fig. 3c). Consequently, the approaches shown in Fig. 3c, d are obviously not perfectly fitted results, since the first 5% of data on the curve could directly affect the HC5 values.

To compare the applicability of the bootstrap and conventional approaches for deriving an HC5 in SSDs, the SSE and RMSE values were calculated (Table 2) (Willmott et al. 1985; Moriasi et al. 2007). The RMSE values in the nonparametric estimates, bootstrap and bootstrap regression, were less than those observed in the parametric estimates. The nonparametric bootstrap approaches (Fig. 3e, f) fit the toxicity data better than the parametric approaches, where various frequency distributions were assumed (Fig. 3a–d).

Relationships between the variation of HC5 and the number of iterations during computational processes of parametric and nonparametric approaches are shown in Fig. 4a–f. The nonparametric processes (Fig. 4e, f) converged more quickly to a sufficiently small value of RSME, after iterations (700) than those conducted by the parametric approaches (2,000) (Fig. 4a–d) (Grist et al. 2002; Wang et al. 2008). Nonparametric approaches were superior (in convergence) to parametric curve-fitting methods during the computational processes (Townsend et al. 2007). In addition, the range of variation in HC5 values estimated by nonparametric methods was also narrower than that generated by parametric approaches. For instance, HC5 values for Zn estimated by the Gompertz distribution ranged 89.4–158.5 µg/L, which was wider than the results conducted by the bootstrap regression with a range of 137.8–167.1 µg/L. Bootstrap methods were generally more stable than those developed by the use of the parametric approaches.

## 3.3 Species Sensitivity

The SSDs used to derive HC5 showed that there was variability in species sensitivity. Compared with other taxa, aquatic plants showed a relatively wide range in sensitivities to all eight metallic elements. For instance, *Chlorella* sp. were generally sensitive to the effects of Cd, Cr, Cu, Hg, and Zn (Fig. 5b–e, h), while they were less sensitive to As and were tolerant to Pb. There was a range in tolerances of the angiosperm, *Lemna minor*. This species showed toxic effects when exposed to Cd and Zn (Fig. 5b, h) but was less sensitive to Cu and Hg (Fig. 5d, e).

Fishes were differentially sensitive to the toxicity of the elements studied. For example, the zebra fish (*Danio rerio*) was sensitive to Hg, Ni, and Pb (Fig. 5f, g), was moderately sensitive to As and Cd (Fig. 5a, b), and was tolerant of Zn (Fig. 5a–h). In contrast, the walking catfish (*Clarias gariepinus*) was tolerant to almost all of the polluting chemicals, especially Cu, Hg, and Ni (Fig. 5d, e, h).

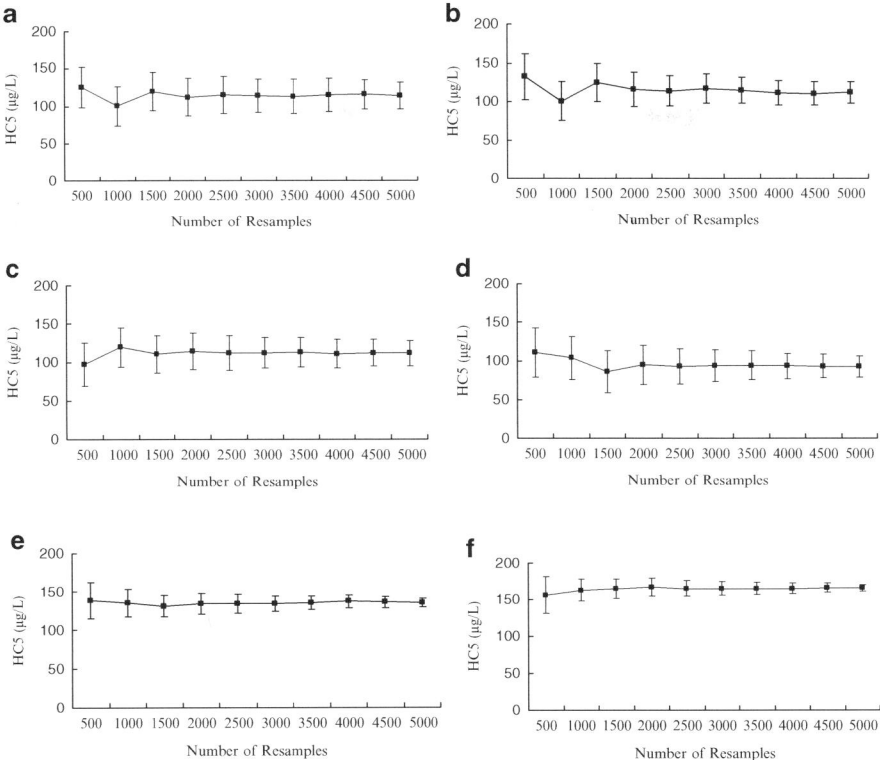

**Fig. 4** Relationship between variation of HC5 of Zn and iterations made by different approaches ($n=49$). (**a**) Gompertz; (**b**) log-logistic; (**c**) lognormal; (**d**) sigmoid; (**e**) bootstrap; (**f**) bootstrap regression. The *solid lines* indicate average HC5 values estimated by different approaches (for 500 iterations of simulation), and the *vertical bars* represent the associated standard deviations

Compared to other taxa, most species of zooplankton were relatively sensitive to the effects of all chemical treatments evaluated. Among all selected species, the model organism, *Daphnia magna*, was the most sensitive to As treatments (Fig. 5a), the third most sensitive to Cd and Cr (Fig. 5b, c) treatments, and was sensitive to Hg and Pb (Fig. 5e, g). This sensitivity of *D. magna* was consistent with the toxic test results in a previous relevant study (OECD 2011).

Macroinvertebrates were sensitive to most selected chemicals among species used to compile the SSDs. For instance, *Gammarus pulex* was the most sensitive species to Pb and the second most sensitive to As (Fig. 5a, g). Mollusks were moderately sensitive to most of the metals and metalloids such as *Mytilus edulis* and *Lamellidens marginalis* exposed to Hg (Fig. 5e) and *Dreissena polymorpha* exposed to Ni (Fig. 5f). Mollusks are suitable for both bio-monitoring and hazard and risk assessment (Borcherding and Volpers 1994; Salánki et al. 2003).

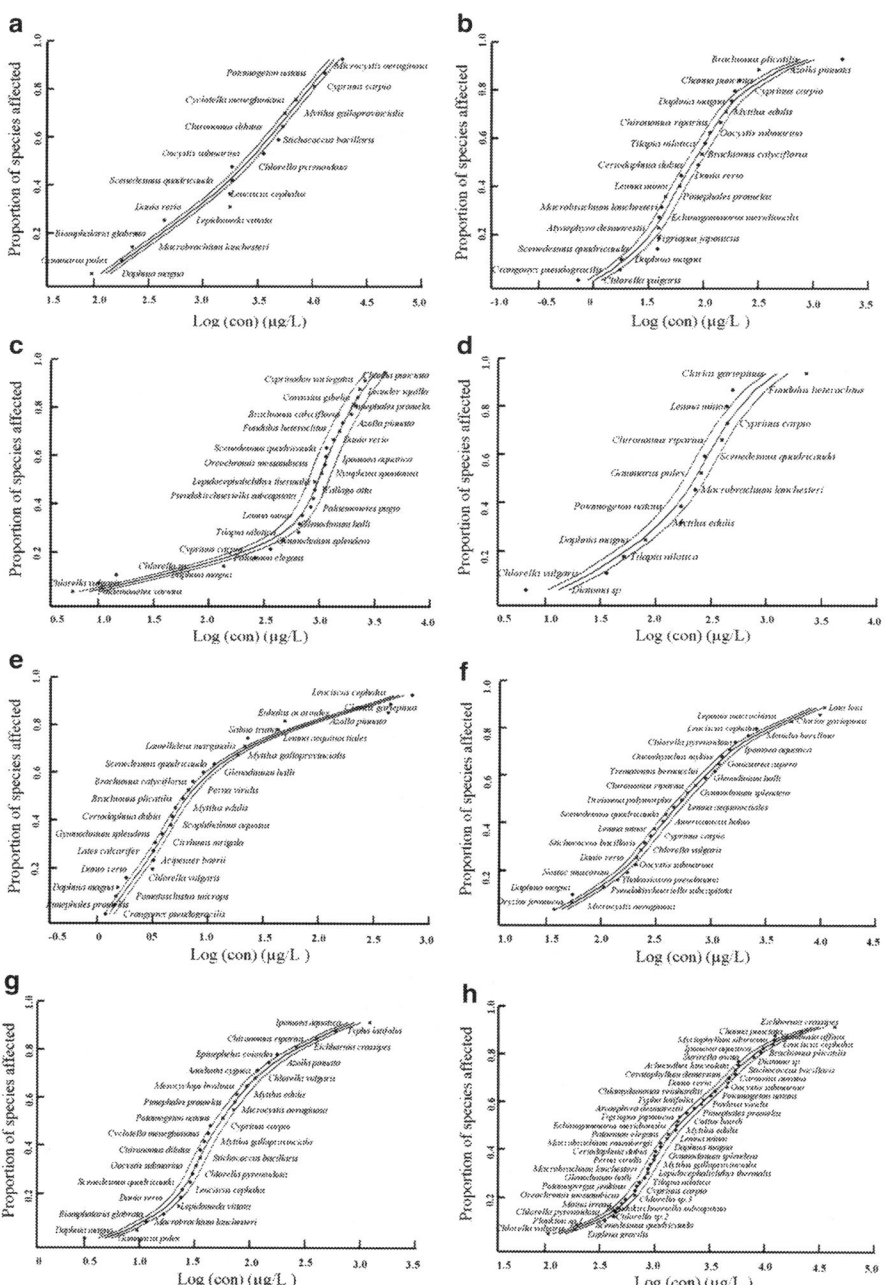

**Fig. 5** SSDs calculated from bootstrap regression with 95% confidence interval and species series ranked by toxicity of metals and metalloids. (**a**) As ($n=17$); (**b**) Cd ($n=22$); (**c**) Cr ($n=27$); (**d**) Cu ($n=14$); (**e**) Hg ($n=26$); (**f**) Ni ($n=29$); (**g**) Pb ($n=28$); (**h**) Zn ($n=49$)

## 4 Risk Assessments

### 4.1 Measured Exposure Concentrations

Arithmetic mean concentrations of eight metals and metalloids to which aquatic species are exposed were measured in the 40 sites of Tai Lake (Table 3). The arithmetic mean values, rather than geometric mean values, were used for risk assessments, since concentrations had little variability (Yin and Fan 2011; Zhang et al. 2012). According to the *China Environmental Quality Standards for Surface Water* (GB3838-2002), the MECs of three metals and metalloids, such as As, Cd, and Hg, were less than the Class I regulation level, and the MECs for the other metals (i.e., Cr, Cu, and Zn) belong to the Class II levels, whereas Pb did not meet the requirements of the Class II level. Compared with the *China Standards for Irrigation Water Quality* (GB5084-2005) and *China Water Quality Standard for Fisheries* (GB11607-89), all metals and metalloids met the requirements except Cu. This indicated that most of the metals and metalloids fulfilled the requirements for employing lake water for uses such as irrigation and fisheries. However, the exposure concentrations of metals and metalloids in Tai Lake were higher than those that existed in a similar lake: Chaohu Lake (Tong et al. 2006). The main reason for this difference was the high industrial and agricultural discharge from Wuxi, Changzhou, and Suzhou that takes place around Tai Lake.

### 4.2 Correlation Analysis

A correlation analysis (Table 4) showed that there was a significant relationship among these metals, and this indicated that they emanated from sources that had

**Table 3** Comparison of values of MEC for metals and metalloids found in Tai Lake, and water quality standards of China from different sources

| Metals and metalloids | MEC (µg/L) | | | Environmental quality standards for surface water (µg/L) | | Standards for irrigation water quality (µg/L) | Water quality standard for fisheries (µg/L) |
|---|---|---|---|---|---|---|---|
| | Range | Arithmetic mean | SD | Class I | Class II | | |
| As | 0.67–12.06 | 4.52 | 1.76 | 50 | 50 | 50 | 50 |
| Cd | 0.76–1.12 | 0.85 | 0.05 | 1 | 5 | 10 | 5 |
| Cr | 31.76–75.50 | 40.04 | 5.6 | 10 | 50 | 100 | 100 |
| Cu | 2.40–170.70 | 18.97 | 21.2 | 10 | 1,000 | 500 | 10 |
| Hg | 0.001–0.246 | 0.0048 | 0.004 | 0.05 | 0.05 | 1 | 0.5 |
| Ni | 16.60–30.91 | 19.61 | 1.86 | – | – | – | 50 |
| Pb | 9.89–29.81 | 16.9 | 3.34 | 10 | 10 | 200 | 50 |
| Zn | 17.66–1,246 | 70.26 | 154.33 | 50 | 1,000 | 2,000 | 100 |

–, no data available; MEC, measured exposure concentration

**Table 4** Correlation coefficients between eight metals and metalloids

|    | As     | Cd     | Cu     | Cr     | Hg     | Ni    | Pb    | Zn    |
|----|--------|--------|--------|--------|--------|-------|-------|-------|
| As | 1.000  |        |        |        |        |       |       |       |
| Cd | 0.134  | 1.000  |        |        |        |       |       |       |
| Cu | 0.005  | 0.301  | 1.000  |        |        |       |       |       |
| Cr | 0.319  | 0.562* | −0.091 | 1.000  |        |       |       |       |
| Hg | −0.072 | −0.276 | −0.062 | −0.183 | 1.000  |       |       |       |
| Ni | 0.292  | 0.448  | −0.032 | 0.714* | −0.399 | 1.000 |       |       |
| Pb | 0.293  | 0.254  | −0.046 | 0.306  | −0.034 | 0.187 | 1.000 |       |
| Zn | 0.264  | 0.624* | −0.068 | 0.873**| −0.151 | 0.544*| 0.306 | 1.000 |

*Significant at the 0.05 level (two tailed)
**Significant at the 0.01 level under the null hypothesis of $\rho=0$

**Table 5** Hazard quotients of metals and metalloids to aquatic species in the Tai Lake calculated by HC5 values derived from SSDs based on six different approaches

| Metals and metalloids | As     | Cd      | Cr      | Cu      | Hg    | Ni     | Pb      | Zn     |
|-----------------------|--------|---------|---------|---------|-------|--------|---------|--------|
| Approaches            |        |         |         |         |       |        |         |        |
| Gompertz              | 0.062  | 4.722** | 5.976** | 3.237** | 0.008 | 0.507* | 7.161** | 0.615* |
| Log-logistic          | 0.107* | 3.696** | 9.059** | 3.221** | 0.009 | 0.544* | 7.578** | 0.628* |
| Lognormal             | 0.056  | 5.313** | 5.317** | 2.769** | 0.009 | 0.682* | 5.748** | 0.677* |
| Sigmoid               | 0.071  | 4.722** | 6.963** | 4.371** | 0.01  | 0.73*  | 8.622** | 0.759* |
| Bootstrap             | 0.0049 | 3.696** | 4.608** | 2.577** | 0.008 | 0.540* | 5.541** | 0.520* |
| Bootstrap regression  | 0.039  | 3.036** | 3.473** | 1.785** | 0.006 | 0.336* | 4.774** | 0.425* |

Asterisk number indicates the risk levels of metals and metalloids: low risk (none), medium risk (*), and high risk (**)

similar and related anthropogenic activities. Rather high correlations were found between Cd and Zn (0.624), Cr and Ni (0.714), and Cr and Zn (0.873). This led us to believe that effluents from neighboring industries and municipal sewage might contribute to substantial loads of metals–metalloids to the rivers flowing from city and rural areas. For example, Cd is usually regarded as deriving from anthropogenic-sourced wastewater and fertilizers, or pesticides, whereas Cr, Ni, and Zn are usually connected with printing or electroplating industry discharges (Li et al. 2009).

## 4.3  Hazard Quotients

The PNEC value is equal to the HC5 value, since the UF was set as 1 in this study (EU 1996). The hazard quotients can be directly derived by dividing the MEC and HC5 (1 and 2). For each HC5 value calculated from SSDs using different approaches, we obtained the corresponding HQ value for assessing the risk of metals and metalloids (Table 5). Generally, the decreasing order of the HQ values for the eight elements was as follows: Hg<As<Ni<Zn<Cu<Cd<Pb<Cr. The more toxic elements, Cr, Pb, and Cd, exhibited greater risks. This is reasonable considering that these three metals have both greater toxic potency and greater rates of discharge. Many factories exist around Tai Lake, including printing houses and electroplating factories, and thereby serve as sources of these metallic contaminants. From our analysis, risks of Cu, Zn, and Ni were somewhat in the middle range, although their human health toxicities are not that great. Of course, these are also essential dietary metals for humans. The last two elements, As and Hg, showed the least risk to aquatic species. This is mainly due to their lesser natural concentrations, although they are commonly thought of as being among the most toxic metals.

Applying different SSD approaches to the data produced different HC5 values for protecting aquatic species from metals and metalloids. Moreover, utilizing different approaches affected the hazard quotients and risk levels for aquatic species. For instance, the HQ for As, calculated by HC5 through bootstrap regression, was 0.039 at the lesser level of risk, whereas it was 0.107 at the medium level of risk when conducted by a log-logistic distribution. Consequently, what model is selected to treat the data is not only important when deriving WQC values but also a key issue when assessing the risk of water pollutants.

## 5 Discussion

### 5.1 *Evaluation of Approaches*

When constructing SSDs, one limitation of conventional parametric methodologies is that no single frequency distribution adequately fits all of the available data (Grist et al. 2006). In particular, the accuracy and precision are poor when sample sizes are very small (Moore and Caux 1997). This effect can be seen from curves fitted with the standard log-logistic and sigmoid distributions for Cr (Fig. 3c). In such cases, the HC5 obtained are distorted. Because bootstrapping does not require designation of a particular statistical distribution of chemical effects on species assemblages, it could be an alternative tool to deal with this limitation. Bootstrapping requires a precondition that the empirical distribution of endpoint values could truly represent the real distribution of the source data in the world (Efron and Tibshirani 1993; Grist et al. 2002). Advantages and disadvantages of both conventional parametric and bootstrap approaches were addressed by Grist et al. (2002), when the bootstrapping regression approach was first introduced to construct SSDs for ecological risk assessments. The bootstrap approach does not force a distribution onto data, but it is a relatively data-intensive technique that completely ignores a priori information about distributions of biological responses. In contrast, the parametric approaches most frequently used in deriving SSDs for ecological risk assessment generally require simple computations but make more demanding assumptions about the distribution of data.

## 5.2 Selection of Approaches

Using toxicity data of representative aquatic species and typical water contaminants in China, a comparison of six approaches showed that nonparametric methods based on bootstrapping were statistically superior to the parametric curve-fitting approaches. These results were generally consistent with previous comparisons of multiple approaches that have been applied to derive hazardous concentrations of contaminants in water. For instance, Wheeler et al. (2002) applied four approaches (viz., log-logistic, lognormal, bootstrap, and bootstrap regression) to construct SSDs based on acute lethality of Ni and Cd to saltwater organisms. They found that curves generated by bootstrap and bootstrap regression best matched toxicity data among the approaches. The superiority of bootstrap methods was also reported when developing SSDs on the toxicity of organochlorine pesticides (Wang et al. 2008). These results suggested that bootstrap methods showed promising applications for protecting sensitive species than conventional parametric fitting methods. However, it is still too early to conclude that the bootstrap is the best model to simulate SSDs for all circumstances. Like HC5 values, the applicability of a particular model could be affected by several factors, including available data amounts, species composition, data selected, chemical toxicity, and geographical characteristics.

As the main components in SSD, the species composition and species sensitivity to chemicals could directly affect modeling of predictive values and accuracy of SSDs. The composition of species and sensitivities of organisms to chemicals in different ecosystems relate to their geographic distribution (Brock et al. 2006). For instance, the most common fishes in China are species of Cyprinidae, while in North America it is Salmonidae. Moreover, because of limited toxicity data from literatures, the species used in this study cannot represent all common aquatic organisms in the natural aquatic environment of China. A more sufficient set of toxicity data, covering as many as possible representative species, will be used in further studies. Applications of the same model and metal on chronic toxicity data would be different from those on acute toxicity data. One example is developing SSDs for acute and chronic toxicity of Zn to Chinese species by using parametric fitting methods (Wu et al. 2011). The exponential distribution ($F(x) = 1 - \exp(-\lambda x)$, $x > 0$) gave the best fit to the acute data, while the sigmoid distribution was superior to other methods for fitting the chronic data.

Considering both the advantages and disadvantages of the approaches investigated in this study, if there are sufficient data and if parametric approaches fit the data well, they should be chosen for because of their computational simplicity (Wheeler et al. 2002). However, if the parametric descriptors fail to fit the toxicity data, in which species number is over than 20, the standard bootstrap methods should be used; otherwise, if the data number is less than 20, the bootstrap regression is a better choice by stochastically inserting values.

## 5.3 Proportion of Species to Be Protected

Values for HC5 derived in this study indicate that if the concentration of a certain pollutant is less than the HC5, more than 95% of aquatic species that are chronically exposed would not be adversely affected (Aldenberg and Jaworska 2000). The species proportion to be protected from pollutant chemicals involved in SSDs would be a key issue for establishing water quality criteria. The goal of the WQC is to ensure that toxicants appearing in water and sediment do not adversely affect all or most populations of species in a particular ecosystem and do not impair the overall structure or function of the ecosystem. Although the use of the fifth centile of the biological species is arbitrary, it is generally applied in slightly to moderately disturbed areas and is widely used. A 99% level of protection is appropriate in areas of greater ecological value or where there are concerns about bioaccumulation or toxicity to endangered species. A lesser level of protection might be appropriate, at least as an interim measure, in more disturbed areas. Consequently, the level adopted varies among countries or geographies. For instance, Canadian guidelines aim to protect 100% of species everywhere from long-term exposure (CCME 1999), whereas Australia (ANZECC and ARMCANZ 2000), the European Union (European Commission 2000), and the United States (USEPA 1986) seek to protect a percentage of species, usually 95%, sometimes 99% (pristine areas) or 80% (heavily modified ecosystems).

## 5.4 HQ for Risk Assessment

HQ values were effectively used for screening-level ecological risk assessments for Tai Lake. However, the results of ecological risk assessment are conservative and preliminary considering several factors such as sampling frequency, available toxicity data, and environmental conditions. First, water samples were only collected over a short period in September of 2009. To better reflect and characterize the status of these metals for the long term, seasonal sampling of metal content is needed. Second, limited numbers of metals and metalloids and exposed species were addressed. Inclusion of additional metals or organic pollutants may change the potential risk to aquatic species. Third, exposure concentrations are dynamic in the context of environmental factors such as temperature, pH, and dissolved oxygen. In addition, the species composition in the assessed target water body is variable from seasonal change. Such dynamic changes in the community need to be considered in future studies.

## 6 Summary

Both nonparametric and parametric approaches were used to construct SSDs for use in ecological risk assessments. Based on toxicity to representative aquatic species and typical water contaminants of metals and metalloids in China, nonparametric methods based on the bootstrap were statistically superior to the parametric curve-fitting approaches. Knowing what the SSDs for each targeted species are might help in selecting efficient indicator species to use for water quality monitoring. The species evaluated herein showed sensitivity variations to different chemical treatments that were used in constructing the SSDs. For example, *D. magna* was more sensitive than most species to most chemical treatments, whereas *D. rerio* was sensitive to Hg and Pb but was tolerant to Zn.

HC5 values, derived for the pollutants in this study for protecting Chinese species, differed from those published by the USEPA. Such differences may result from differences in geographical conditions and biota between China and the United States. Thus, the degree of protection desired for aquatic organisms should be formulated to fit local conditions. For approach selection, we recommend all approaches be considered and the most suitable approaches chosen. The selection should be based on the practical information needs of the researcher (viz., species composition, species sensitivity, and geological characteristics of aquatic habitats), since risk assessments usually are focused on certain substances, species, or monitoring sites.

We used Tai Lake as a typical freshwater lake in China to assess the risk of metals and metalloids to the aquatic species. We calculated hazard quotients for the metals and metalloids that were found in the water of this lake. Results indicated the decreasing ecological risk of these contaminants in the following order: Hg<As<Ni<Zn<Cu<Cd<Pb<Cr. From the methodological perspective, six SSD approaches used delivered different WQC values and affected the risk assessment results of the metals and metalloids to aquatic species. Based on the MEC and HC5 derived from SSDs by nonparametric and parametric approaches together, the risk levels of metals and metalloids were characterized from their hazard quotients as being high risk (Cr, Pb, Cd, and Cu), medium risk (Zn and Ni), or low risk (As and Hg).

**Acknowledgements** The research was supported by the National Basic Research Program of China (973 Program) (No. 2008CB418200) and the National Natural Science Foundation of China (No. U0833603 and 41130743). Prof. Giesy was supported by the program of 2012 "High Level Foreign Experts" (#GDW20123200120) funded by the State Administration of Foreign Experts Affairs, the P.R. China to Nanjing University, and the Einstein Professor Program of the Chinese Academy of Sciences. He was also supported by the Canada Research Chair program, a Visiting Distinguished Professorship in the Department of Biology and Chemistry and State Key Laboratory in Marine Pollution, City University of Hong Kong. The authors give special thanks to Dr. Yanping Chen in Southern Medical University, China, for his helpful comments on solving statistical problems during the review process.

# References

Aldenberg T, Jaworska JS (2000) Uncertainty of the hazardous concentration and fraction affected for normal species sensitivity distributions. Ecotoxicol Environ Saf 46:1–18

Aldenberg T, Slob W (1993) Confidence limits for hazardous concentrations based on logistically distributed NOEC toxicity data. Ecotoxicol Environ Saf 25:48–63

ANZECC, ARMCANZ (2000) Australian and New Zealand guidelines for fresh and marine water quality. Australian and New Zealand Environment and Conservation Council and Agriculture and Resource Management Council of Australia and New Zealand, Canberra

Balk F, Okkerman PC, Dogger JW (1995) Guidance document for aquatic effects assessment. Organization for Economic Co-operation and Development (OECD), Paris

Batley GE (2012) "Heavy metal"—a useful term. Integr Environ Assess Manag 8:215

Borcherding J, Volpers M (1994) The "Dreissena-monitor". First results on the application of this biological early warning system in the continuous monitoring of water quality. Water Sci Technol 29:199–201

Brock TCM, Arts GHP, Maltby L, Van den Brink PJ (2006) Aquatic risks of pesticides, ecological protection goals, and common aims in European Union legislation. Integr Environ Assess Manag 2:20–46

Calow P (1996) Variability: noise or information in ecotoxicology? Environ Toxicol Pharmacol 2:121–123

Campbel P, Stokes P (1985) Acidification and toxicity of metals to aquatic biota. Can J Fish Aquat Sci 42:2034–2049

Campbell KR, Bartell SM, Shaw JL (2000) Characterizing aquatic ecological risks from pesticides using a diquat dibromide case study. II. Approaches using quotients and distributions. Environ Toxicol Chem 19:760–774

CCME (1999) Canadian environmental quality guidelines. Canadian Council of Ministers of the Environment, Winnipeg

Chapman PM (2012) "Heavy metal"—cacophony, not symphony. Integr Environ Assess Manag 8:216

R Development Core Team (2011) R: a language and environment for statistical computing. R Foundation for Statistical Computing, Vienna, Austria. ISBN 3-900051-07-0, URL http://www.R-project.org/

Duffus JH (2002) "Heavy metals" a meaningless term? Pure Appl Chem 74:793–807

ECB (2003) Technical guidance document on risk assessment—Part II. Technical report. Institute for Health and Consumer Protection, Ispra, Italy

Efron B, Tibshirani R (1993) An introduction to the bootstrap. Chapman & Hall, Boca Raton, FL

EU (1996) Technical guidance document on risk assessment in support of council directive 93/67/EEC on risk assessment for new notified substances, commission regulation (EC) 1488/94 on risk assessment for existing substances

European Commission (2000) Directive 2000/60/EC of the European Parliament and of the Council establishing a framework for the Community action in the field of water policy. OJEC L327:1–72

Forbes TL, Forbes VE (1993) A critique of the use of distribution-based extrapolation models in ecotoxicology. Funct Ecol 7:249–254

Giesy JP, Solomon KR, Coats JR, Dixon KR, Giddings JM, Kenaga EE (1999) Chlorpyrifos: ecological risk assessment in North American aquatic environments. Rev Environ Contam Toxicol 160:1–129

Grist EPM, Leung KMY, Wheeler JR, Crane M (2002) Better bootstrap estimation of hazardous concentration thresholds for aquatic assemblages. Environ Toxicol Chem 21:1515–1524

Grist EPM, O'Hagan A, Crane M, Sorokin N, Sims I, Whitehouse P (2006) Bayesian and time-independent species sensitivity distributions for risk assessment of chemicals. Environ Sci Technol 40:395–401

Jagoe R, Newman MC (1997) Bootstrap estimation of community NOEC values. Ecotoxicology 6:293–306

Klimisch H, Andreae M, Tillmann U (1997) A systematic approach for evaluating the quality of experimental toxicological and ecotoxicological data. Regul Toxicol Pharmacol 25:1–5

Li JL, He M, Han W, Gu YF (2009) Analysis and assessment on heavy metal sources in the coastal soils developed from alluvial deposits using multivariate statistical methods. J Hazard Mater 164:976–981

Mance G (1987) Pollution threat of heavy metals in aquatic environments. Springer, New York

Marchant C, Leiva V, Cavieres MF, Sanhueza A (2013) Statistical distributions with application to PM10 in Santiago, Chile. Rev Environ Contam Toxicol 223:1–31

Meng W, Wu FC (2010) Introduction of water quality criteria theory and methodology (in Chinese). Science Press, Beijing

Moore DRJ, Caux PY (1997) Estimating low toxic effects. Environ Toxicol Chem 16:794–801

Moriasi D, Arnold J, Van Liew M, Bingner R, Harmel R, Veith T (2007) Model evaluation guidelines for systematic quantification of accuracy in watershed simulations. Trans ASABE 50:885–900

Newman MC, Ownby DR, Mezin LCA, Powell DC, Christensen TRL, Lerberg SB, Anderson BA (2000) Applying species sensitivity distributions in ecological risk assessment: assumptions of distribution type and sufficient numbers of species. Environ Toxicol Chem 19:508–515

Nogawa K, Kido T (1993) Biological monitoring of cadmium exposure in itai-itai disease epidemiology. Int Arch Occup Environ Health 65:43–46

OECD (2011) Proposal for updated guideline 211 *Daphnia magna* reproduction test. Organization for Economic Co-operation and Development, Paris

Pennington DW (2003) Extrapolating ecotoxicological measures from small data sets. Ecotoxicol Environ Saf 56:238–250

Posthuma L, Suter GW, Traas TP (2002) Species sensitivity distributions in ecotoxicology. CRC Press, Boca Raton, FL

Power M, McCarty LS (1997) Fallacies in ecological risk assessment practices. Environ Sci Technol 31:370A–375A

Rausina GA, Wong DCL, Raymon Arnold W, Mancini ER, Steen AE (2002) Toxicity of methyl tert-butyl ether to marine organisms: ambient water quality criteria calculation. Chemosphere 47:525–534

RIVM (2007) Guidance document on deriving environmental risk limits in The Netherlands, report 601501012. National Institute of Public Health and the Environment, Bilthoven, The Netherlands

Salánki J, Farkas A, Kamardina T, Rózsa KS (2003) Molluscs in biological monitoring of water quality. Toxicol Lett 140:403–410

Sanchez-Bayo F, Baskaran S, Kennedy IR (2002) Ecological relative risk (EcoRR): another approach for risk assessment of pesticides in agriculture. Agric Ecosyst Environ 91:37–57

Shao Q (2000) Estimation for hazardous concentrations based on NOEC toxicity data: an alternative approach. Environmetrics 11:583–595

Shaw WHR, Grushkin B (1957) The toxicity of metal ions to aquatic organisms. Arch Biochem Biophys 67:447–452

Solomon KR, Baker DB, Richards RP, Dixon KR, Klaine SJ, La Point TW, Kendall RJ, Weisskopf CP, Giddings JM, Giesy JP, Hall LW, Williams WM (1996) Ecological risk assessment of atrazine in North American surface waters. Environ Toxicol Chem 15:31–76

Solomon KR, Giesy JP, Jones P (2000) Probabilistic risk assessment of agrochemicals in the environment. Crop Prot 19:649–655

SPSS Inc. (2008) SPSS statistics for windows, version 17.0. SPSS Inc., Chicago, IL

Stephan CE, Mount DI, Hansen DJ, Gentile JH, Chapman GA, Brungs WA (1985) Guidelines for deriving numeric national water quality criteria for the protection of aquatic organisms and their uses. PB85-227049. National Technical Information Services, Springfield, VA

TenBrook PL, Palumbo AJ, Fojut TL, Hann P, Karkoski J, Tjeerdema RS (2010) The University of California-Davis methodology for deriving aquatic life pesticide water quality criteria. In: Whitacre DM (ed) Reviews of environmental contamination and toxicology, vol 209. Springer Science+Business Media, LLC, New York, pp 1–155

Tong JH, Huang XM, Chen Y (2006) Evaluation of contamination of heavy metal in Chaohu lake (in Chinese). J Anhui Agric Sci 17:4373–4374

Townsend PA, Papeş M, Eaton M (2007) Transferability and model evaluation in ecological niche modeling: a comparison of GARP and Maxent. Ecography 30:550–560

Tsushima K, Naito W, Kamo M (2010) Assessing ecological risk of zinc in Japan using organism- and population-level species sensitivity distributions. Chemosphere 80:563–569

USEPA (1968) Report of the subcommittee of water quality criteria. Technical report. US Department of the Interior, Washington, DC

USEPA (1976) Quality criteria for water. EPA 440/9-76/023. Office of Water Planning and Standards, Washington, DC
USEPA (1985) Guidelines for deriving numerical national water quality criteria for the protection of aquatic organisms and their uses, PB-85-227049. US Environmental Protection Agency, National Technical Information Service, Springfield, VA
USEPA (1986) Quality criteria for water. EPA-440/5-86-001. Office of Water Regulations and Standards, Washington, DC
USEPA (1996) Water quality criteria documents for the protection of aquatic life in ambient water: 1995 updates. EPA-820-B-96-001. Office of Water, Washington, DC
USEPA (1999) National recommended water quality criteria-correction. EPA-822-Z-99-001. Office of Water, Washington, DC
USEPA (2002) National recommended water-quality criteria: 2002. EPA-822-R-02-047. Office of Water, Washington, DC
USEPA (2004) National recommended water quality criteria. Office of Water, Office of Science and Technology, Washington, DC
USEPA (2009) National recommended water quality criteria. Office of Water, Washington, DC
van Dam RA, Harford AJ, Warne MSJ (2012) Time to get off the fence: the need for definitive international guidance on statistical analysis of ecotoxicity data. Integr Environ Assess Manag 8:242–245
van Sprang PA, Verdonck FAM, van Assche F, Regoli L, De Schamphelaere KAC (2009) Environmental risk assessment of zinc in European freshwaters: a critical appraisal. Sci Total Environ 407:5373–5391
van Straalen NM (2002) Threshold models for species sensitivity distributions applied to aquatic risk assessment for zinc. Environ Toxicol Pharmacol 11:167–172
van Straalen NM, Denneman CAJ (1989) Ecotoxicological evaluation of soil quality criteria. Ecotoxicol Environ Saf 18:241–251
Vardy DW, Tompsett AR, Sigurdson JL, Doering JA, Zhang X, Giesy JP, Hecker M (2011) Effects of subchronic exposure of early life stages of white sturgeon (*Acipenser transmontanus*) to copper, cadmium, and zinc. Environ Toxicol Chem 30:2497–2505
Varian H (2005) Bootstrap tutorial. Math J 9:768–775
Wagner C, Løkke H (1991) Estimation of ecotoxicological protection levels from NOEC toxicity data. Water Res 25:1237–1242
Wang B, Yu G, Huang J, Hu H (2008) Development of species sensitivity distributions and estimation of HC5 of organochlorine pesticides with five statistical approaches. Ecotoxicology 17:716–724
Wheeler JR, Grist EPM, Leung KMY, Morritt D, Crane M (2002) Species sensitivity distributions: data and model choice. Mar Pollut Bull 45:192–202
WHO (2006) Guidelines for drinking water quality: incorporation 1st and 2nd addenda. Technical report, vol 1, 3rd edn. WHO, Geneva, recommendations
Willmott CJ, Ackleson SG, Davis RE, Feddema JJ, Klink KM, Legates DR, O'Donnell J, Rowe CM (1985) Statistics for the evaluation and comparison of models. J Geophys Res 90:8995–9005
Wu FC, Meng W, Zhao XL, Li HX, Zhang RQ, Cao YJ, Liao HQ (2010) China embarking on development of its own national water quality criteria system. Environ Sci Technol 44:7992–7993
Wu FC, Feng CL, Cao YJ, Zhang RQ, Li HX, Liao HQ, Zhao XL (2011) Toxicity characteristic of zinc to freshwater biota and its water quality criteria (in Chinese). Asian J Ecotoxicol 6:367–382
Wu FC, Feng CL, Zhang RQ, Li YS, Du DY (2012) Derivation of water quality criteria for representative water-body pollutants in China. Sci China Earth Sci 55:900–906
Yin HY, Fan C (2011) Dynamics of reactive sulfide and its control on metal bioavailability and toxicity in metal-polluted sediments from Lake Taihu, China. Arch Environ Contam Toxicol 60:565–575
Zhang Y, Hu X, Yu T (2012) Distribution and risk assessment of metals in sediments from Taihu Lake, China using multivariate statistics and multiple tools. Bull Environ Contam Toxicol 89:1009–1015

# Toxicity Reference Values for Protecting Aquatic Birds in China from the Effects of Polychlorinated Biphenyls

Hailei Su, Fengchang Wu, Ruiqing Zhang, Xiaoli Zhao, Yunsong Mu, Chenglian Feng, and John P. Giesy

## Contents

| | | |
|---|---|---|
| 1 | Introduction | 60 |
| 2 | Data Collection and Analysis Methods | 61 |
| | 2.1 Selection of Representative Species in China and Toxicity Data | 61 |
| | 2.2 Methods for Deriving TRVs and TRGs | 62 |
| 3 | Review of PCB Bird Toxicity Studies | 64 |
| | 3.1 Domestic Chicken (Gallus gallus domesticus) | 64 |
| | 3.2 Double-Crested Cormorant (Phalacrocorax auritus) | 68 |
| | 3.3 Common Tern (Sterna hirundo) | 69 |
| | 3.4 Osprey (Pandion haliaetus) | 69 |
| | 3.5 Bald Eagle (Haliaeetus leucocephalus) | 69 |
| | 3.6 American Kestrel (Falco sparverius) | 70 |
| | 3.7 Great Horned Owl (Bubo virginianus) | 70 |
| 4 | Derivation of TRVs and TRGs | 70 |
| | 4.1 Species Sensitivity Distribution Method | 70 |
| | 4.2 Critical Study Approach | 73 |
| | 4.3 Toxicity Percentile Rank Method | 74 |
| 5 | Results and Discussion | 75 |

H. Su
College of Water Sciences, Beijing Normal University, Beijing 100875, China

State Key Laboratory of Environmental Criteria and Risk Assessment,
Chinese Research Academy of Environmental Sciences, Beijing 100012, China

F. Wu (✉) • R. Zhang • X. Zhao • Y. Mu • C. Feng
State Key Laboratory of Environmental Criteria and Risk Assessment,
Chinese Research Academy of Environmental Sciences, Beijing 100012, China
e-mail: wufengchang@vip.skleg.cn

J.P. Giesy
Department of Veterinary Biomedical Sciences and Toxicology Centre,
University of Saskatchewan, Saskatoon, SK, Canada

Zoology Department and Center for Integrative Toxicology, Michigan State University,
East Lansing, MI 48824, USA

| | | |
|---|---|---|
| 6 | Assessment of the Risk PCBs Pose to Birds | 76 |
| | 6.1 Comparison of TRVs to PCB Concentrations in Birds | 76 |
| | 6.2 Comparison of TRGs to PCB Concentrations in Fish | 77 |
| 7 | Summary | 77 |
| References | | 78 |

# 1 Introduction

Polychlorinated biphenyls (PCBs) are widely distributed, persistent, bioaccumulative, and toxic pollutants of abiotic matrices, such as soils, sediments, and water, and of wildlife such as fish and birds (Kannan et al. 2000; Giesy et al. 1994b). PCBs are stable in water and adsorb to particles that can be deposited in sediment or accumulated in aquatic food webs. Because of their persistent and lipophilic properties, PCBs bioaccumulate and biomagnify through aquatic food webs, and if thresholds for adverse effects are exceeded, cause effects on wildlife. As top predators that feed at the top of the aquatic food chain, fish-eating birds are exposed to greater concentrations of PCBs (Giesy et al. 1994a; Bosveld and Van den Berg 1994). Potential adverse effects reported for PCBs on wild birds include reduced hatchability, embryonic deformities, immune suppression, mortality (Kannan et al. 2000; Giesy et al. 1994a; Bosveld and Van den Berg 1994; Brunström 1989; Bosveld et al. 1995, 2000; Hoffman et al. 1998), and population-level effects (CCME 1998; Sanderson et al. 1994).

PCBs cause dioxin-like effects by binding to the aryl hydrocarbon receptor (AhR) in birds (Safe 1990, 1994; Kennedy et al. 1996). The primary PCB congeners that cause AhR-mediated effects to birds are the non- and mono-*ortho*-substituted congeners (Bosveld and Van den Berg 1994; Bosveld et al. 2000). As the most sensitive biological effect, ethoxyresorufin-*O*-deethylase (EROD) activity induction has been suggested to be a suitable biochemical indicator for exposure to dioxin-like compounds, such as some PCB congeners (Elliott et al. 2001). It is thought that the developing embryo is the life stage that is most sensitive to the toxic effects of pollutants (Lam et al. 2008; Peterson et al. 1993). Accordingly, PCB concentrations in eggs were the more predictive measure of exposure to derive toxic reference values (TRVs). The tissue-based risk of great blue herons was assessed on the basis of an egg-based TRV, and was developed by taking the geometric mean of effect concentrations in three egg-injection studies (Seston et al. 2010).

PCB congeners have different toxic potencies because their physical and chemical properties and structures are different. Such differences determine the binding affinity of PCBs to the AhR. Assessments of the PCB hazard to humans and wildlife are complicated because PCBs occur in mixtures that change as a function of time from weathering, differential accumulation, and metabolism. Log-$K_{ow}$ values for the PCBs are between 4.3 and 8.26. PCBs that have even moderate $K_{ow}$ values accumulate in aquatic organisms and their predators. Relative Potency Factors (RePs) have been used to calculate concentrations of 2,3,7,8-TCDD toxicity equivalents (TEQ) in samples as the sum of the product of the ReP multiplied by the concentration of the respective congeners (Van den Berg et al. 1998).

Wildlife, such as birds, consumes persistent substances mainly via consumption of fish, crustaceans, invertebrates, and plants. Governments have established allowable residue concentrations in tissues to protect wildlife that feed on aquatic organisms contaminated by persistent, bioaccumulative, and toxic substances. Tissue residue guidelines (TRGs) are concentrations of xenobiotics in tissues that are established for aquatic biota to protect wildlife that consume them. Generally, as concentrations of xenobiotics increase, the greater is the expectation that adverse effects will occur on birds and mammals feeding on aquatic organisms (CCME 1998). The USEPA (1995a) used NOAEL (no observed adverse effects levels) or LOAEL (lowest observed adverse effects levels) values to derive wildlife criteria for PCBs. Hatching success of pheasant eggs was the endpoint and appropriate uncertainty factors were used. In addition, the EPA derived wildlife criteria for the PCBs for kingfisher, silver gulls, and bald eagles, and the geometric mean of the values for these three species was taken as the PCB wildlife criterion for birds by the USEPA. The tissue residue guideline for PCBs in aquatic birds derived in Canada was 2.4 ng TEQs/kg food wet mass (wm), and was based on a toxicity study in white Leghorn hens (CCME 1998, 2001).

We had two objectives in the present study:

1. Derive TRVs and TRGs for the effects of PCBs on aquatic birds by using the toxicity percentile rank method (TPRM), the species sensitivity distribution (SSD) and critical study approach (CSA), along with the method used in the USA and Canada for deriving the wildlife criteria for PCBs in birds.
2. Assess the PCB hazard to birds by comparing TRVs and TRGs derived in this study, with the actual concentrations of PCBs measured in birds and fish in selected regions of China. The additional value contributed by this study is that it provides a scientific baseline for risk management of PCBs in China.

## 2 Data Collection and Analysis Methods

### 2.1 Selection of Representative Species in China and Toxicity Data

The primary criterion for selecting representative avian species is their exposure to pollutants via aquatic food webs (USEPA 1995a), such as fish-eating birds. The night heron (*Nycticorax nycticorax*), little egret (*Egretta garzetta*), and Eurasian spoonbill (*Platalea leucorodia*) were selected as three representative avian species in China (ZRQ). All of these are widely distributed in Chinese aquatic ecosystems and are known to feed on aquatic prey (Barter et al. 2005). These three species have been studied extensively as bioindicators of wetland health and environmental pollution (Lam et al. 2008; Levengood et al. 2007; An et al. 2006). Body masses (bm) and rates of food ingestion (FI) for these three avian species are shown in Table 1.

**Table 1** Body masses (bm) and food ingestion (FI) rates of several avian birds

| Avian species | bm (kg) | FI (kg/day wm[a]) | FI:bm | Reference |
|---|---|---|---|---|
| Night heron | 0.706 | 0.239 | 0.34 | Zhang et al. (2013) |
| Little egret | 0.342 | 0.148 | 0.43 | Zhang et al. (2013) |
| Eurasian spoonbill | 2.232 | 0.514 | 0.23 | Zhang et al. (2013) |
| Common tern | 0.127 | 0.0774 | 0.61 | Nagy (2001) |
| Chicken | 2.0 | 0.134 | 0.067 | USEPA (1995a) |
| Ring-necked pheasant | 1.1 | 0.0638 | 0.058 | USEPA (1995a) |
| Japanese quail | 0.12 | 0.012 | 0.10 | USEPA (1995a) |
| Northern bobwhite | 0.04 | 0.0072 | 0.18 | USEPA (2007) |
| Mourning dove | 0.128 | 0.058[b] | 0.45 | Nelson and Martin (1953) |
| Ring dove | 0.149 | 0.0169 | 0.11 | USEPA (2007) |
| American kestrel | 0.12 | 0.0444 | 0.37 | USEPA (1995a) |
| Screech owl | 0.194 | 0.02 | 0.10 | USEPA (2007) |
| Mallard | 1.082 | 0.25 | 0.23 | CCME (1998) |

[a] wm stands for wet mass
[b] Calculated from the allometric equation (Nagy 2001): $FI = 2.065 \times bm^{0.689}$

The effects of PCBs on birds have been reviewed and summarized (USEPA 1995a; Barron et al. 1995; Bosveld and Van den Berg 1994), and toxicity threshold values for PCBs have been derived from NOAEL or LOAEL levels established for several toxicity endpoints. Toxicity data for dietary exposure were converted to tolerable daily intake (TDI) values, which were calculated from bm and food ingestion rates of selected surrogate birds. Utilizable NOAEL or LOAEL values were selected, based on the principles given in the following document: "Protocol for the derivation of Canadian tissue residue guidelines for the protection of wildlife that consume aquatic biota" (CCME 1998). The main principals followed are as follows: (1) studies were constructed under suitable control conditions and considered ecological-relevant endpoints, such as reproduction and embryonic development; (2) only chronic or subchronic studies with a clear dose–response relationship were accepted; (3) the form and dosage of tested chemicals were reported in the study.

## 2.2 Methods for Deriving TRVs and TRGs

PCBs occur in the environment as weathered mixtures, and weathered residue profiles differ from those of the original technical mixtures. Therefore, assessment of hazards posed by PCBs to wildlife must account for changes in the relative proportions of PCB congeners and their different toxic potencies. Accordingly, the concept of Relative Potency Factors (ReP) was introduced to allow comparisons of the toxicity of a compound relative to TCDD, based on its available in vivo and in vitro data (Van den Berg et al. 1998). In this approach, it is assumed that the combined effects of different congeners were either dose- or concentration-additive. Concentrations of 2,3,7,8-TCDD equivalents (TEQs) of PCBs can be calculated by using toxic equivalency factors (TEFs) and available chemical residue data (1). Application of an NOAEL or LOAEL value as a reference dose could either be overprotective or under-protective and may not reflect the specific dose–response relationship (Kannan et al. 2000). To address this problem, the geometric mean of the NOAEL and LOAEL values are used as the reference concentration (RC) (Kannan et al. 2000) (2). If the NOAEL value was not determined in a particular study, it can be estimated by dividing the LOAEL by a factor of 5.6 (CCME 1998) (3). The tolerable daily intake (TDI) is calculated as shown in (4).

$$TEQ = \sum (PCB_i \times TEF_i) \qquad (1)$$

$$RC = (NOAEL \times LOAEL)^{0.5} \qquad (2)$$

$$NOAEL = LOAEL / 5.6 \qquad (3)$$

$$TDI = RC \times (FI / bm) \qquad (4)$$

Three methods used to derive TRGs and TRVs are (1) Species sensitivity distribution (SSD), (2) Critical study approach (CSA), and (3) Toxicity percentile rank method (TPRM). Each of the three has advantages and disadvantages. A species sensitivity distribution (SSD) is a probability distribution function that can be used to describe the range of tolerances among species (Leo Posthuma et al. 2002). The SSD method has been used widely in aquatic ecological risk assessment and derivation of water quality criteria (WQC) for aquatic biota (Caldwell et al. 2008; Hall et al. 2009). The SSD makes full use of available toxicity data and represents the whole ecosystem. But this approach is not often applied when assessing risks to wildlife, because so little toxicity data for wildlife are available. We used the SSD method in this study to derive the TRVs and TRGs of PCBs for protection of fish-eating birds, and we used the most sensitive endpoint data for each species (USEPA 2005). The SSD approach assumes that sensitivities of species can be described by a specified statistical distribution (e.g., normal distribution). If the selected toxic data for PCBs can be described by using a log-normal distribution, then the ETX2.0 program can be employed to fit the distribution. Calculating an $HC_5$ (Hazard Concentration affecting 5% of species) via this program gives a value that protects 95% species from contaminants. Moreover, it provides two-sided 90% confidence limits designated as an upper limit (UL $HC_5$) and lower limit (LL $HC_5$) (Zhang et al. 2013).

The critical study approach (CSA) is a primary method used for risk assessment and criteria derivation for wildlife (Kannan et al. 2000; CCME 2001; USEPA 1995a; Sample et al. 1993; Newsted et al. 2005). The CSA method has the advantage of requiring less data and being simpler to calculate. This method depends mainly on the toxicity values of sensitive species and has greater uncertainty. Results from available toxicity studies on the targeted species were selected in this method as the basis for deriving TRVs (Blankenship et al. 2008). TRVs for wildlife were then calculated by using the lowest toxicity value from the critical study (tissue level or dietary concentration); appropriate uncertainty factors (UFs) were also applied.

Uncertainty factors were determined primarily from guidance given by the US-EPA (USEPA 1995a, b; Weseloh et al. 1995). Three types of uncertainty factors were considered: interspecies uncertainty factor ($UF_A$), sub-chronic to chronic uncertainty factor ($UF_S$), and LOAEL-to-NOAEL uncertainty factor ($UF_L$). Values of 1–10 were assigned to represent the degree of uncertainty for each factor and were based on the nature of available scientific information as well as professional judgment.

The toxicity percentile rank method (TPRM) is the standard method recommended by the USEPA for deriving water quality criteria for protecting aquatic organisms (USEPA 1985). The TPRM more comprehensively reflects the toxic effects of pollutants to organisms and ultimately provides better protection for wildlife. When using the TPRM, the reference concentrations (RC) for avian species are first ordered from largest to least, and ranks ($R$) are assigned to RCs from 1 for the lowest to $N$ ($N$ is the number of avian species) for the highest. Second, the cumulative probability $P$ is calculated for each species using the equation: $P = R/(N+1)$. Finally, four RCs, which have cumulative probabilities closest to 0.05 (always the four least RCs,) are selected as the basis to calculate the TRVs (5–8).

$$S^2 = \frac{\sum\left[(\ln RC)^2\right] - \left[\sum(\ln RC)\right]^2 / 4}{\sum(P) - \left[\sum(\sqrt{P})\right]^2 / 4} \qquad (5)$$

$$L = \left\{\sum(\ln RC) - S\left[\sum(\sqrt{P})\right]\right\} / 4 \qquad (6)$$

$$A = S(\sqrt{0.05}) + L \qquad (7)$$

$$TRV = e^A \qquad (8)$$

## 3 Review of PCB Bird Toxicity Studies

The toxicity of PCBs to birds, emphasizing reproduction and developmental effects, was summarized by Barron et al. (1995). To augment the information from Barron et al. (1995), additional recent and relevant toxicity studies of PCB effects on birds were compiled, reviewed, and critiqued. All available toxicity data (both diet and tissue data) were summarized and are presented in Table 2.

### 3.1 *Domestic Chicken* (Gallus gallus domesticus)

It has been shown in several studies that chickens are among the most sensitive species to the effects of PCBs; moreover, PCB126 was the most toxic congener and

Table 2 A summary of the toxicity of PCB isomers and mixtures to birds (both for tissue and diet)

| Species | PCBs | Toxicity end point | NOAEL[a] | LOAEL[b] | References |
|---|---|---|---|---|---|
| Chicken (tissue) | PCB126 | Egg mortality | | 0.2 ng/g wm | Powell et al. (1996) |
| | PCB126 | EROD[c] activity | | 0.3 ng/g wm | Hoffman et al. (1998) |
| | PCB77 | EROD activity | 0.12 ng/g wm | 1.2 ng/g wm | Hoffman et al. (1998) |
| | PCB126 | Egg mortality | | 1.0 ng/g wm | McKernan et al. (2007) |
| | PCB126 | Thymocyte apoptosis | 0.13 ng/g wm | 0.32 ng/g wm | Goff et al. (2005) |
| | PCB126 | Immune function | | 0.25 ng/g egg | Lavoie and Grasman (2007) |
| | PCB126 | Egg mortality | 0.051 ng/g wm | 0.128 ng/g wm | Fox and Grasman (1999) |
| | PCB1254, 1242 | MDI[d] activity | | 6.7 mg/kg wm | Gould et al. (1999) |
| | PCBs | Hatching success | 0.95 mg/kg wm | 1.5 mg/kg wm | Britton and Huston (1973) |
| | | Hatching success | 0.36 mg/kg wm | 2.5 mg/kg wm | Scott (1977) |
| | | Egg production | | 5 mg/kg wm | Platonow and Reinhart (1973) |
| | | Hatching success | | 4 mg/kg wm | Tumasonis et al. (1973) |
| Chicken (diet) | Aroclor 1016, 1221, 1254 | Reproductive efficiency | 20 mg/kg food | | Lillie et al. (1974, 1975) |
| | Aroclor 1232, 1268 | Egg production | | 20 mg/kg food | Lillie et al. (1974) |
| | Aroclor 1232, 1242, 1248 | Hatching success | 5 mg/kg food | 10 mg/kg food | Britton and Huston (1973), Lillie et al. (1975) |
| | Aroclor 1242 | Hatching success | 2 mg/kg food | 20 mg/kg food | Lillie et al. (1974) |
| | Aroclor 1248, 1254 | Chick growth | | 2 mg/kg food | Lillie et al. (1974) |
| | Aroclor 1248 | Hatching success | 1 mg/kg food | 10 mg/kg food | Scott (1977) |
| | Aroclor 1254 | Egg production | | 5 mg/kg food | Platonow and Reinhart (1973) |
| | Aroclor 1254 | Hatching success | | 50 mg/kg food | Tumasonis et al. (1973) |
| Double-crested cormorant (tissue) | PCB126 | Egg mortality | | 25 ng/g wm | Powell et al. (1997) |
| | PCB126 | EROD induction | | 70 ng/g wm | Powell et al. (1998) |
| | PCBs | Egg mortality | | 3.5 mg/kg wm | Tillitt et al. (1992) |

(continued)

Table 2 (continued)

| Species | PCBs | Toxicity end point | NOAEL[a] | LOAEL[b] | References |
|---|---|---|---|---|---|
| American kestrel (tissue) | PCB126 | EROD induction | 23 ng/g wm | 233 ng/g wm | Hoffman et al. (1998) |
|  | PCB77 | EROD induction | 100 ng/g wm | 1000 ng/g wm | Hoffman et al. (1998) |
|  | PCB126 | EROD induction |  | 50 ng/g wm | Hoffman et al. (1996) |
| American kestrel (diet) | Aroclor 1254 | Male fertility |  | 33 mg/kg food | Bird et al. (1983) |
|  | Aroclor 1254 | Female fertility | 0.5 mg/kg food | 5 mg/kg food | Linger and Peakall (1970) |
| Common tern (tissue) | PCBs | EROD induction | 25ng TEQ's/g lipid |  | Bosveld et al. (2000) |
|  | PCB126 | EROD induction |  | 44 ng/g wm | Hoffman et al. (1998) |
|  | PCBs | Reproductive success | 7 mg/kg wm | 8 mg/kg wm | Bosveld and Van den Berg (1994) |
|  |  | Reproductive success | 5 mg/kg wm |  | Hoffman et al. (1993) |
|  |  | Reproductive success | 4 ng TEQs/g lipid |  | Bosveld and Van den Berg (1994) |
| Common tern (diet) | PCBs | EROD induction | 0.6 ng TEQs/g food |  | Bosveld et al. (2000) |
| Caspian tern (tissue) | PCBs | Reproductive success |  | 4.2 mg/kg wm | Yamashita et al. (1993) |
|  | PCBs | Egg shell thickness | 30 mg/kg wm |  | Struger and Weseloh (1985) |
| Bald eagle (tissue) | PCBs | EROD induction | 0.1 ng TEQs/g wm | 0.21 ng TEQs/g wm | Elliott John et al. (1996) |
|  | PCBs | Reproductive success |  | 4 mg/kg wm | Wiemeyer et al. (1984) |
|  |  | Reproductive success |  | 13 mg/kg wm | Bosveld and Van den Berg (1994) |
| Osprey (tissue) | PCBs | EROD induction | 0.037 ng TEQs/g wm | 0.13 ng TEQs/g wm | Elliott et al. (2001) |
| Herring gull (tissue) | PCBs | Reproductive success |  | 5 mg/kg wm | Ludwig et al. (1993) |
| Great horned owl (tissue) | PCBs | EROD induction | 0.14 ng TEQs/g wm | 0.4 ng TEQs/g wm | Strause et al. (2007) |
| Great blue heron (tissue) | PCBs | Reproductive success | 7.8 mg/kg wm |  | Boily et al. (1994) |

| Species (matrix) | Chemical | Endpoint | NOAEL[a] | LOAEL[b] | Reference |
|---|---|---|---|---|---|
| Black-crowned night heron (tissue) | PCBs | Breeding success | 10.9 mg/kg wm | | Tremblay and Ellison (1980) |
| Mallard (tissue) | Aroclor 1254 | Reproductive success | 23.3 mg/kg wm | | Custer and Heinz (1980) |
| | Aroclor 1242 | Egg shell thinning | | 105 mg/kg wm | Haseltine and Prouty (1980) |
| Mallard (diet) | Aroclor 1254 | Reproductive success | 25 mg/kg food | | Custer and Heinz (1980) |
| | Aroclor 1242 | Egg shell thing | | 150 mg/kg food | Haseltine and Prouty (1980) |
| Screech owl (tissue) | Aroclor 1248 | Reproductive success | 7.1 mg/kg wm | | Anne et al. (1980) |
| Screech owl (diet) | Aroclor 1248 | Reproductive success | 3 mg/kg food | | Anne et al. (1980) |
| Forster's tern (tissue) | PCBs | Reproductive success | 7 mg/kg wm | 19 mg/kg wm | Bosveld and Van den Berg (1994) |
| | PCBs | Hatching success | 4.5 mg/kg wm | 22.2 mg/kg wm | Kubiak et al. (1989) |
| | | Hatching success | 0.2 ng TEQs/g wm | 2.2 ng TEQs/g wm | Kubiak et al. (1989) |
| Ringed turtle dove (tissue) | Aroclor 1254 | Hatching success | | 16 mg/kg wm | Peakall et al. (1972) |
| Ringed turtle dove (diet) | Aroclor 1254 | Hatching success | | 10 mg/kg food | Peakall et al. (1972), Peakall and Peakall (1973) |
| | Aroclor 1254 | Brain neurotransmitter concentrations | 1 mg/kg food | 10 mg/kg food | Heinz et al. (1980) |
| Mourning dove (diet) | Aroclor 1254 | Reproductive behavior | | 10 mg/kg food | Tori and Peterle (1983) |
| Northern bobwhite (diet) | Aroclor 1254 | Reproductive effects | 50 mg/kg food | | Eisler (1986) |
| Japanese quail (diet) | Aroclor 1254 | Egg shell thickness | 50 mg/kg food | | Chang and Stokstad (1975) |
| | Aroclor 1248 | Egg shell quality | 20 mg/kg food | | Scott (1977) |
| Ring-necked pheasant (diet) | Aroclor 1254 | Female fertility | | 50 mg/kg food | Roberts et al. (1978) |

[a] No observed adverse effects level
[b] Lowest observed adverse effects level
[c] Ethoxyresorufin-*O*-deethylase
[d] Monodeiodinase
[e] Toxicity equivalents

the major contributor to TEQs (Wiesmüller et al. 1999; Senthilkumar et al. 2002; Strause et al. 2007).

The toxic effects of PCB126 and PCB77 on chickens (*Gallus gallus*) through hatching were studied, focusing on embryonic development and induction of EROD activity (Hoffman et al. 1998). Doses of two congeners were injected into chicken embryos at the following doses: PCB126 (0.3, 0.5, 1, or 3.2 ng/g wet mass (wm)), or PCB77 (0.12, 1.2, 6, or 12 ng/g wm). The $LD_{50}$ and LOAEL values of PCB126 for chicken were 0.4 and 0.3 ng/g wm and the $LD_{50}$, NOAEL, and LOAEL of PCB77 were 2.6, 0.12, and 1.2 ng/g wm, respectively. In addition, the effects of PCB126 (0.1, 0.2, 0.4, 0.8, 1.6, 3.2, 6.4, or 12.8 ng PCB126/g wm egg) on hatching and development of chicken were also investigated via injection into eggs. The LOAEL of PCB126 was 0.2 ng/g wm (Powell et al. 1996). Three doses (viz., 0.5, 1, and 2.0 ng PCB126/g wm egg) were injected into eggs, and the LOAEL value, based on hatching success, survival, and edema, was 1.0 ng PCB126/g wm egg (McKernan et al. 2007). The toxic effects of PCB126 on chicken embryos were determined by injection of 0.051, 0.13, 0.32, or 0.80 ng PCB126/g egg, wm. The NOEC and LOEC values (0.051 and 0.13 ng/g wm egg, respectively) for PCB126 (injected into chicken embryos) were based on mortality, immune organ quality, and lymphocyte structure (Fox and Grasman 1999).

The effects of PCB126 on thymus atrophy in chicken embryo was examined by egg injection (0.05, 0.13, 0.32, 0.64, or 0.80 ng/g wm egg); the resulting $LD_{50}$, NOEC, and LOEC values were respectively 1.01, 0.13, or 0.32 ng/g wm egg (Goff et al. 2005). The effects of PCB126 on death and immune function of chickens were determined by egg injection, and the LOAEL was 0.25 ng/g wm egg (Lavoie and Grasman 2007). LOAELs of total PCBs for reproductive success of chicken were 1,500–5,000 ng/g of egg wm (Barron et al. 1995), which were greater than PCB126 values (the most toxic congener of PCBs).

## 3.2 Double-Crested Cormorant (Phalacrocorax auritus)

The double-crested cormorant is a typical fish-eating bird that has exhibited adverse effects attributed to PCB exposure. Such effects include embryonic lethal and developmental defects, from ingesting fish contaminated with PCBs and other substances. Based on a toxic endpoint of egg mortality, the LOAEL value for PCBs was 3,500 ng/g wm egg (Tillitt et al. 1992; Yamashita et al. 1993). In another study, eggs of double-crested cormorants were injected with 5, 10, 25, 50, 100, 200, 400, or 800 ng PCB126/g wm to examine double-crested cormorant toxicity. The $LD_{50}$ and LOAEL values of PCB126 for double-crested cormorant were 158 and 25 ng/g egg, respectively (Powell et al. 1997). The $LD_{50}$ of PCB126 was greater than that for chicken, which value was 2.3 ng/g wm egg. This result indicated that chickens were more sensitive to the effects of PCB126 than were double-crested cormorants (Powell et al. 1997; Fox and Grasman 1999). In another study, the co-planar PCB congener, PCB126, was injected into the eggs of double-crested cormorants at doses of 60, 150, 300, or 600 ng/g egg. Based on induction of EROD activity, the LOAEL and $LD_{50}$ values of PCB126 for double-crested cormorant were 70 and 177 ng/g wm egg, respectively (Powell et al. 1998).

### 3.3 Common Tern (Sterna hirundo)

As fish-eating birds that top the aquatic food chain, common terns are exposed to the lipophilic and persistent PCBs (Bosveld et al. 1995). The toxicity of PCB126 on common tern was examined by egg injection at three doses (0, 240 or 434 ng/g wm). The $LD_{50}$ and LOAEL, based on reproductive success and induction of EROD activity, were 104 and 44 ng/g wm, respectively (Hoffman et al. 1998).

Biochemical and reproductive effects of PCB126 on chicks of common terns fed fish containing PCB126, or mixtures of PCB126 and PCB153, were examined (Bosveld et al. 2000). The most sensitive parameter affected by PCBs was induction of EROD activity. A nonlinear concentration–effect relationship was observed between TEQ concentrations and EROD activity induction. The LOAEL value for induction of EROD activity was 25 ng TEQs/g liver lipid mass (lm), and was caused by 0.6 ng TEQs/g wm food. The lipid content of common tern was assumed to be 1.9% (Ricklefs 1979), and the LOAEL value was calculated to be 475 pg TEQs/g wm. Based on reproductive success, the NOAEL of PCBs for common tern was <4 ng TEQs/g lm (Bosveld and Van den Berg 1994).

### 3.4 Osprey (Pandion haliaetus)

As a top predator, osprey accumulates lipophilic substances and accordingly can be used as a biological indicator of exposure to PCBs in the aquatic ecosystems (Elliott et al. 2001). Elliott et al. (2001) studied the ecological effects of PCBs on osprey chicks and found that the NOAEL and LOAEL values, based on induction of EROD activity, were 37 and 130 ng TEQs/kg wm, respectively (Elliott et al. 2001).

### 3.5 Bald Eagle (Haliaeetus leucocephalus)

As a top predator of the aquatic food chain, bald eagles feed mainly on fish and other fish-eating birds (Knight et al. 1990). PCBs can reduce the reproductive success of bald eagle populations (Anthony et al. 1993; Wiemeyer et al. 1993). Chicks of bald eagles were more sensitive to the effects of PCBs than were adults. Induction of EROD activity was the most sensitive biomarker for bald eagles exposed to PCBs (Sanderson et al. 1994). Based on induction of EROD activity, the LOAEL and NOAEL values were 100 and 210 ng TEQs/kg wm, respectively (Elliott John et al. 1996).

## 3.6 American Kestrel (Falco sparverius)

Exposure of the American kestrel to a daily intake of 7 mg PCB/kg bm/day produced a body concentration of 34.1 mg PCBs/kg wm, which was consistent with PCB concentrations in birds collected from the Great Lakes basin (Fernie et al. 2001b; Fisher et al. 2006). Breeding behavior was affected and reproductive success was reduced. Subsequent studies have shown that embryos exposed to PCBs could affect propagation of offspring (Fernie et al. 2001a). The developmental toxicity of American kestrel was studied by egg injection of PCB126 (0, 2.3, 23, or 233 ng/g wm) or PCB77 (0, 100, or 1,000 ng/g wm) (Hoffman et al. 1998). Results showed that the LOAELs, based on induction of EROD activity, were 233 and 1,000 ng/g wm for PCB126 and PCB77, respectively. When American kestrel were exposed orally to 50, 250, or 1,000 ng/g bm, the LOAEL value, based on developmental toxicity, was 50 ng PCB126/g bm (Hoffman et al. 1996).

## 3.7 Great Horned Owl (Bubo virginianus)

Great horned owls are another species that feeds at the top of the terrestrial food chain, and hence are very sensitive to PCBs, and are often used as a biological indicator species. Based on induction of EROD activity, the NOAEL and LOAEL values for PCB exposure were 135 and 400 pg TEQs/g wm egg, respectively (Strause et al. 2007; Elliott John et al. 1996).

## 4 Derivation of TRVs and TRGs

TRGs and TRVs for birds were derived for PCBs by using three approaches: SSD, TPRM, and CSA. The toxic endpoints recorded for PCBs on birds are shown in Table 2. Because of the toxicity differences among PCB congeners, TEQs were selected and used to derive TRVs and TRGs. NOAEL and LOAEL values for the most sensitive toxicity endpoints were selected and the geometric mean of these two values was used as the reference concentration (RC). If a value was available only for the NOAEL or the LOAEL, then the other one was calculated by using (3). Toxicity data were transformed to equivalent concentrations by using TEFs for birds (Van den Berg et al. 1998), and TEFs for PCB commercial mixtures (Table 3).

## 4.1 Species Sensitivity Distribution Method

Using the data selection principles mentioned above, toxicity data (tissue) on PCBs was selected to derive a TRV of PCBs for the following birds: chicken,

**Table 3** Toxic equivalent conversion factors for some commercial PCB mixtures for birds (CCME 2001)

| Mixture | Conversion factor (ng TEQ/mg product) |
|---|---|
| Aroclor 1242 | 234.6 |
| Aroclor 1248 | 251.3 |
| Aroclor 1254 | 44.5 |
| Aroclor 1260 | 25.5 |

**Table 4** Reference concentrations (RC) used to construct SSD curves (pg TEQs/g wm)

| Avian species | PCBs[a] NOAEL/LOAEL | TEQs[b] NOAEL/LOAEL | RC[c] | Reference |
|---|---|---|---|---|
| Chicken | 0.051/0.128 | 0.0051/0.0128 | 8 | Fox and Grasman (1999) |
| Double-crested cormorant | 0.45/25 | 0.45/2.5 | 1,060 | Powell et al. (1997) |
| American kestrel | 8.9/50 | 0.89/5 | 2,110 | Hoffman et al. (1996) |
| Common tern | 7.9/44 | 0.79/4.4 | 1,860 | Hoffman et al. (1998) |
| Bald eagle | N/N | 0.1/0.21 | 140 | Elliott John et al. (1996) |
| Osprey | N/N | 0.037/0.13 | 70 | Elliott et al. (2001) |
| Great horned owl | N/N | 0.14/0.4 | 240 | Strause et al. (2007) |
| Mallard | 23,300/130,480 | 1.04/5.82 | 2,460 | Custer and Heinz (1980) |
| Screech owl | 7,100/39,760 | 1.78/9.97 | 4,210 | Anne et al. (1980) |
| Forster's tern | 4,500/22,200 | 0.2/2.2 | 660 | Kubiak et al. (1989) |
| Ringed turtle dove | 2,857/16,000 | 0.13/0.71 | 90 | Peakall et al. (1972) |

*N* no data
[a] ng PCBs/g egg
[b] ng TEQs/g wm egg
[c] pg TEQs/g bm/day

double-crested cormorant, American kestrel, common tern, bald eagle, osprey, great horned owl, mallard, screech owl, Forster's tern, and ringed turtle dove (Table 4). From the data on these species, the SSD was constructed and the hazard concentration affecting 5% of species ($HC_5$) was estimated by using the ETX2.0 software. The $HC_5$ value, which theoretically protects 95% species, is known to be a good predictor of the threshold for community-level effects. We determined the $HC_5$ value, along with the 90% upper and lower confidence limits (UL $HC_5$ and LL $HC_5$) and show them in Fig. 1. The $HC_5$ was predicted to be 15.5 pg TEQs/g, wm, which was defined as the TRV with UL $HC_5$ and LL $HC_5$ of 1.8 and 54.5 pg TEQs/g, wm, respectively.

Toxicity data on the PCBs for common tern, chicken, ring-necked pheasant, Japanese quail, Northern bobwhite, mourning dove, ringed turtle dove, American kestrel, screech owl, and mallard based on diet were selected to derive the TRG of PCBs (Table 5). Data expressed as TEQs were obtained by use of TEFs (1). The Total Daily Intake (TDI) was calculated from FI and bm using (2) and (4) (Table 1). TDI values were calculated for all ten avian species as shown in Table 5.

The TDI values were then used to construct a SSD curve (Fig. 2). The $HC_5$ value was predicted to be 3.43 pg TEQs/kg bm/day, and the UL $HC_5$ and LL $HC_5$ were

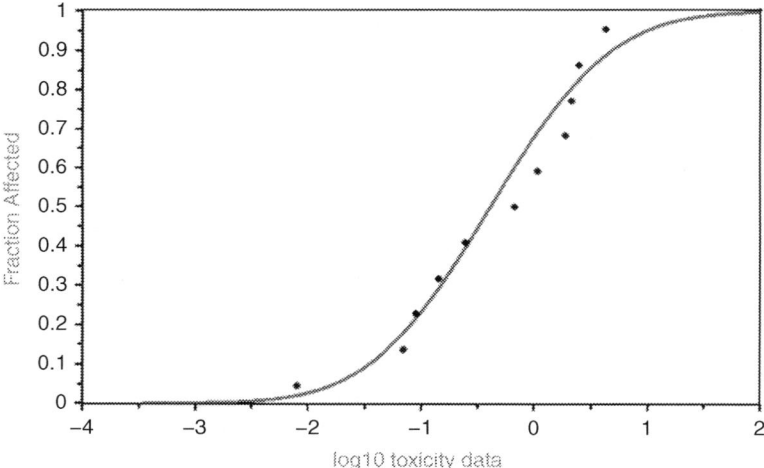

**Fig. 1** Distribution of species sensitivity (SSD) for toxicity of PCBs to birds (pg TEQs/g wm). The $HC_5$ was 15.5 pg TEQs/g wm, and UL $HC_5$ and LL $HC_5$ were 1.8 and 54.5 pg TEQs/g wm, respectively

**Table 5** The toxicity data values, based on dietary exposure, that were used to fit the SSD curve

| Avian species | PCBs[a] NOAEL/ LOAEL | TEQs[b] NOAEL/ LOAEL | TDI[c] | Reference |
|---|---|---|---|---|
| Common tern | N/N | 0.11/0.6 | 158.6 | Bosveld et al. (2000) |
| Chicken | 0.36/2 | 0.016/0.089 | 2.6 | Lillie et al. (1974) |
| Ring-necked pheasant | 8.9/50 | 0.4/2.2 | 54.5 | Roberts et al. (1978) |
| Japanese quail | 50/280 | 2.2/12.3 | 520 | Chang and Stokstad (1975) |
| Northern bobwhite | 50/280 | 2.2/12.3 | 936 | Eisler (1986) |
| Mourning dove | 1.8/10 | 0.08/0.45 | 85.5 | Tori and Peterle (1983) |
| Ringed turtle dove | 1/10 | 0.045/0.45 | 154 | Heinz et al. (1980) |
| American kestrel | 0.5/5 | 0.022/0.22 | 25.9 | Linger and Peakall (1970) |
| Screech owl | 3/16.8 | 0.13/0.73 | 9.5 | Anne et al. (1980) |
| Mallard | 25/140 | 1.1/6.2 | 598 | Custer and Heinz (1980) |

*TDI* tolerable daily intake, *N* no data
[a] mg PCBs/kg wm food
[b] ng TEQs/g wm food
[c] pg TEQs/g bm/day

respectively 0.35 and 12.5 pg TEQs/kg bm/day. Using FI/bm values of three representative avian species listed in Table 1, RCs for these three bird species were calculated to be 10.1, 7.98, and 14.9 pg TEQs/g food. The geometric mean of these three RCs was 10.7 pg TEQs/g food, which was defined as the TRG value in birds for PCBs.

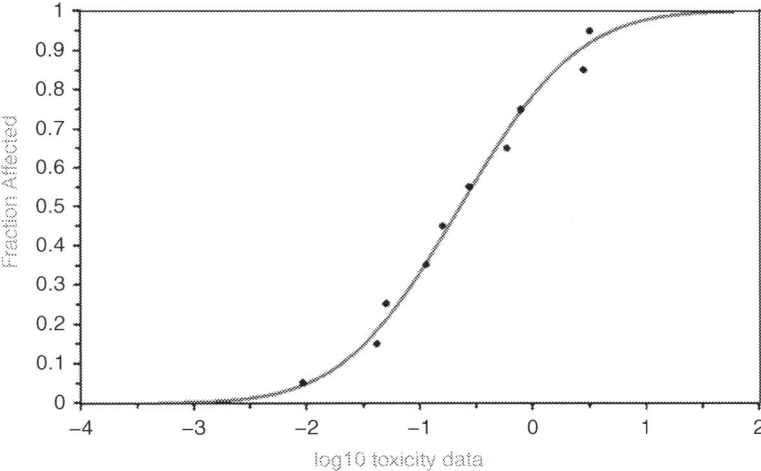

**Fig. 2** Distribution of species sensitivity for avian toxicity data of PCBs based on diet exposure. The HC$_5$ was 3.43 pg TEQs/g/day, and UL HC$_5$ and LL HC$_5$ were 0.35 and 12.5 pg TEQs/g bm/day, respectively

## 4.2 Critical Study Approach

Based on induction of EROD activity as the toxicity endpoint, the NOAEL and LOAEL values for PCBs on osprey were 37 and 130 pg TEQs/g wm, respectively (Elliott et al. 2001). The geometric mean of these two values (69.4 pg TEQs/g wm) was taken as the RC. Similarly, based on induction of EROD activity the NOAEL and LOAEL values for PCBs on bald eagle were 100 and 210 pg TEQs/g wm, respectively (Elliott John et al. 1996), and the RC was 144.9 pg TEQs/g wm. Osprey and bald eagle are two representative birds at the top of the aquatic food web, and both of these species are sensitive to the effects of PCBs. Induction of EROD activity is the most sensitive biological effect, and is a suitable biochemical indicator for exposure to PCBs in birds (Bosveld and Van den Berg 1994; Bosveld et al. 2000). EROD induction activity is the critical endpoint that occurs at the least exposure concentration. Although conservative, application of this assessment endpoint should be protective of population-level adverse effects. By using the toxicity data available for osprey and bald eagle, the TRV for PCB bird effects can be derived for protecting aquatic birds (CCME 1998; Newsted et al. 2005). The UF$_A$, UF$_L$, and UF$_S$ were set to be 2, 1, and 3, respectively, and the total UF was 6. The TRV of PCBs for birds was then calculated to be 16.7 pg TEQs/g wm, by dividing the geometric mean of two RCs for these birds by a total uncertainty factor of 6.

The common tern is a piscivorous bird at the top of aquatic food chain that can accumulate PCBs in their tissues and eggs. Based on induction of EROD activity as the toxicity endpoint, the LOAEL value for PCBs on common tern was 0.6 ng

**Table 6** The RCs and relevant values used to calculate TRV[a] for TPRM[b] (pg TEQs/g)

| Rank | Avian species | RC | ln RC | (ln RC)$^2$ | $P = R/(N+1)$ | $P^{0.5}$ |
|---|---|---|---|---|---|---|
| 4 | Bald eagle | 140 | 4.94 | 24.42 | 0.33 | 0.58 |
| 3 | Ringed turtle dove | 90 | 4.50 | 20.25 | 0.25 | 0.50 |
| 2 | Osprey | 70 | 4.25 | 18.05 | 0.17 | 0.41 |
| 1 | Chicken | 8 | 2.08 | 4.32 | 0.08 | 0.29 |
| Sum | | | 15.77 | 67.04 | 0.83 | 1.77 |

[a] Toxic reference value
[b] Toxicity percentile rank method

**Table 7** The TDIs and associated relevant values used to calculate TRG[a] for TPRM (pg TEQs/kg/day)

| Rank | Avian species | TDI | ln TDI | (ln TDI)$^2$ | $P = R/(N+1)$ | $P^{0.5}$ |
|---|---|---|---|---|---|---|
| 4 | Ring-necked pheasant | 54.5 | 4.00 | 16.00 | 0.36 | 0.60 |
| 3 | American kestrel | 25.9 | 3.25 | 10.59 | 0.27 | 0.52 |
| 2 | Screech owl | 9.50 | 2.25 | 5.07 | 0.18 | 0.43 |
| 1 | Chicken | 2.60 | 0.96 | 0.91 | 0.09 | 0.30 |
| Sum | | | 10.46 | 32.57 | 0.91 | 1.85 |

[a] Tissue residue guideline

TEQs/g wm (Bosveld et al. 2000). This study was taken as the critical study for deriving a TRG value for PCBs in birds. By employing a total uncertain factor of 6, the calculated TRG value was 42.3 pg/g wm food.

## 4.3 Toxicity Percentile Rank Method

Four of the lowest RC values for bald eagle, ringed turtle dove, osprey and chicken were selected to calculate the TRV for birds by using the toxicity centile rank method. $R = 1, 2, 3, 4$, and $N = 11$. The relevant values are given in Table 6. Based on the values in Table 6 and equations (5–8), the calculated results were $S = 10.2$, $L = -0.57$, $A = 1.71$, and TRV = RC = 5.5 pg TEQs/g wm.

Four of the lowest TDI values for ring-necked pheasant, American kestrel, screech owl, and chicken were selected to calculate TRG for birds by using the toxicity centile rank method. $R = 1, 2, 3, 4$, and $N = 10$. The relevant values are presented in Table 7. Based on the values in Table 7 and equations (5–8), the calculated results were $S = 9.82$, $L = -1.93$, $A = 0.27$, and TDI = 1.31 pg TEQs/kg/day. Using the values of FI:bm for three representative birds (0.23, 0.34, 0.43) in China, the respective RC values were calculated to be 4.53, 3.05, and 5.70 pg TEQs/g wm food. The geometric mean of these three RCs was 4.3 pg TEQs/g wm food, and this value represents the TRG for birds.

## 5 Results and Discussion

PCBs, which are persistent, bioaccumulative, and toxic are widely distributed in the environment. Apical predators at the top of the aquatic food web, such as fish-eating birds are exposed to greater concentrations of PCBs than are primary and secondary producers. TRGs and TRVs for PCBs in birds derived in this study by SSD, CSA, and TPRM were 10.7, 42.3, 4.3 pg TEQs/g diet wm, and 15.5, 16.7, 5.5 pg TEQs/g tissue wm, respectively (see Table 8). The values derived by the three methods had certain differences, which may have resulted from differences in the toxicity data and calculation methods used. The values of TRGs and TRV derived by using the TPRM were smaller than those determined by applying the other two methods. Because the TPRM used the four lowest toxicity data values for the most sensitive species, the criterion calculated was small and might be over-protective for avian species. However, the values derived from all three methods were similar. The CSA has greater uncertainty because it relies on fewer studies and the uncertainty factor for it is based on judgment and experience. When deriving TRG and TRVs for PCBs in birds by using the SSD approach, PCB toxicity data on about ten avian species were employed. Therefore, the TRG of 10.7 pg TEQs/g diet wm and TRV of 15.5 pg TEQs/g tissue wm, derived by using the SSD method, were recommended as criteria for protecting aquatic birds in China from PCBs.

A TRG of 2.4 pg TEQs/g food wm for PCBs was developed in Canada for protecting avian species that consume aquatic biota (CCME 2001). This TRG was based on a PCB toxicity study in white leghorn chickens, which is one of the most sensitive avian species to PCBs (Barron et al. 1995). As a result of this sensitivity, this TRG would likely be overprotective for wild birds. The TRG value for PCBs in birds derived in this study was 10.7 pg TEQs/g wm food, which is slightly greater than 2.4 pg TEQs/g food wm. Body masses and rates of ingestion of food for three representative avian species were used to derive the PCB TRG values for China. Thus, the TRGs derived by using SSD and TPRM were regarded to be more reasonable than the Canadian TRG for performing risk assessments of PCBs on wild birds in China.

The TRV for effects of PCBs on birds, based on concentrations in tissues (including eggs) developed in this study, was 15.5 pg TEQs/g wm. This value was slightly higher than the TRVs for $TEQ_{WHO-Avian}$ (0.8–2.9 pg/g wm) that were used to assess ecological risk of great horned owls exposed to PCDD/DF (Coefield et al. 2010).

However, the PCB toxicity data for birds were limited, and uncertainties existed in deriving TRGs and TRVs for PCBs in birds. Food web structure and environmental factors affect the exposure and effects of birds to PCBs. Therefore, further research into the potential for toxic effects of PCBs on birds in China is needed. In addition, more studies of the structure of food webs for avian species in China are needed.

Table 8 The TRGs and TRVs of PCBs for birds by three methods

| Methods | SSD | CSA | TPRM |
|---|---|---|---|
| Tissue (pg TEQs/g) | 15.5 | 16.7 | 5.5 |
| Diet (pg TEQs/g food) | 10.7 | 42.3 | 4.3 |

## 6 Assessment of the Risk PCBs Pose to Birds

### 6.1 Comparison of TRVs to PCB Concentrations in Birds

As top predators, aquatic birds can accumulate high concentrations of persistent organic compounds, such as PCBs and thus, are often used as receptors of concern in ecological risk assessments. The embryo is the most sensitive life stage for a number of pollutants. Concentrations of pollutants reach young birds primarily via the diet of the female, and PCB concentrations in bird bodies have been found to not correlate with age. PCB concentrations that have been detected in birds from various areas of the world have been summarized in Table 9.

PCB concentrations detected in egrets collected from southern China were approximately 900–3,800 ng/g lm, and the TEQs of PCBs in birds from Hong Kong were greatest (Lam et al. 2008). PCB concentrations in eggs of egrets and black crown night herons from Hong Kong contained levels of 960 (270–1,700) ng/g wm and 230 (85–600) ng/g wm, respectively (Connell et al. 2003).

PCBs in black crown night herons from Chicago contained 586.4–4,678.9 ng/g wm, with an average level of 2,229.6 ng/g wm (Levengood and Schaeffer 2010). The TEQs for the PCBs were not given in these studies. For avian species, PCB concentrations were highest in common tern from the Netherlands and Belgium, followed by cormorants from Japan.

The TEQs for PCBs in birds from Michigan were 1 pg/g wm in eastern bluebirds to 247 pg/g wm in tree swallows, which were comparatively less than those in birds from Japan (17 pg/g wm in whimbrels to 691 pg/g wm in gray herons) (Table 9). TEQs of PCBs in common terns from the Netherlands and Belgium had the greatest level (997 pg/g wm). Based on tissue concentrations, the TRVs of PCBs derived in this study were 5.5–16.7 pg TEQs/g, wm. Most PCB concentrations in birds were

**Table 9** Actual PCB concentrations detected in birds from different geographic zones

| Zones | Avian species | PCBs (ng/g) | TEQs (pg/g) | Reference |
|---|---|---|---|---|
| Michigan | House wren | 24 | 10 | Fredricks et al. (2010) |
| | Tree swallow | 110 | 247 | Fredricks et al. (2010) |
| | Eastern bluebird | 8 | 1 | Fredricks et al. (2010) |
| | Great blue heron | 223 | 130 | Seston et al. (2010) |
| Japan | Cormorant | 8,327 | 409 | Guruge et al. (2000) |
| | Gray heron | 30 | 691 | Senthilkumar et al. (2002) |
| | Spot-billed duck | 20 | 37 | Senthilkumar et al. (2002) |
| | Whimbrel | 30 | 17 | Senthilkumar et al. (2002) |
| | Short-tailed shearwater | 3 | 24 | Senthilkumar et al. (2002) |
| | Cattle egret | 342 | 266 | Senthilkumar et al. 2002) |
| | Great egret | 504 | 134 | Senthilkumar et al. (2002) |
| The Netherlands and Belgium | Common tern | 43,586 | 997 | Bosveld et al. (1995) |

greater than these TRVs. Therefore, it may be that some species of wild birds are experiencing harmful effects from PCB exposure, which is consistent with reported incidents of PCB effects on birds.

## 6.2 *Comparison of TRGs to PCB Concentrations in Fish*

PCB concentrations were measured in 20 species of fish (i.e., ten each from fresh and marine waters) from aquatic environments of the Pearl River Delta in China (Wei et al. 2011). Results showed that levels of PCBs in fish ranged from 0.065 to 5.25 pg TEQ/g wm. Based on the levels of PCBs in fish sampled from the Hudson River and New York Bight (Hong and Bush 1990), the TEQs were calculated to be from 0.47 to 6.86 ng TEQs/g wm, which were much higher than those in China.

The TRGs of PCBs calculated in this study were 4.3–42.3 pg TEQs/g food wm, which were higher than most PCB concentrations in fish from the Pearl River Delta in China (0.065–5.25 pg TEQ/g wm). It was indicated that food consumption would not cause harmful effects to birds. But the PCB concentrations in fish from the Hudson River and New York Bight were higher than the TRGs derived in this study, which showed that harmful effects would be caused to birds from food exposure.

## 7 Summary

PCBs are typical of persistent, bioaccumulative and toxic compounds (PBTs) that are widely distributed in the environment and can biomagnify through aquatic food webs, because of their stability and lipophilic properties. Fish-eating birds are top predators in the aquatic food chain and may suffer adverse effects from exposure to PCB concentrations.

In this review, we address the toxicity of PCBs to birds and have derived tissue residue guidelines (TRGs) and toxic reference values (TRVs) for PCBs for protecting birds in China. In deriving these protective indices, we utilized available data and three approaches, to wit: species sensitivity distribution (SSD), critical study approach (CSA) and toxicity percentile rank method (TPRM). The TRGs and TRVs arrived at by using these methods were 42.3, 10.7, 4.3 pg TEQs/g diet wm and 16.7, 15.5, and 5.5 pg TEQs/g tissue wm for the CSA SSD and TPRM approaches, respectively. These criteria values were analyzed and compared with those derived by others. The following TRG and TRV, derived by SSD, were recommended as avian criteria for protecting avian species in China: 10.7 pg TEQs/g diet wm and 15.5 pg TEQs/g tissue wm, respectively. The hazard of PCBs to birds was assessed by comparing the TRVs and TRGs derived in this study with actual PCB concentrations detected in birds or fish.

The criteria values derived in this study can be used to evaluate the risk of PCBs to birds in China, and to provide indices that are more reasonable for protecting

Chinese avian species. However, several sources of uncertainty exists when deriving TRGs and TRVs for the PCBs in birds, such as lack of adequate toxicity data for birds and need to use uncertainty factors. Clearly, relevant work on PCBs and birds in China are needed in the future. For example, PCB toxicity data for resident avian species in China are needed. In addition, studies are needed on the actual PCB levels in birds and fish in China. Such information is needed to serve as a more firm foundation for future risk assessments.

**Acknowledgements** This work was supported by National Basic Research Program of China (2008CB418200) and National Natural Science Foundation of China (41261140337, 40973090).

# References

An W, Hu J, Wan Y, An L, Zhang Z (2006) Deriving site-specific 2,2-bis(chlorophenyl)-1, 1-dichloroethylene quality criteria of water and sediment for protection of common tern populations in Bohai bay, north China. Environ Sci Technol 40(8):2511–2516

Anne M, McLane R, Hughes DL (1980) Reproductive success of screech owls fed Aroclor® 1248. Arch Environ Contam Toxicol 9(6):661–665

Anthony RG, Garrett MG, Schuler CA (1993) Environmental contaminants in bald eagles in the Columbia River estuary. J Wildl Manage 57:10–19

Barron M, Galbraith H, Beltman D (1995) Comparative reproductive and developmental toxicology of PCBs in birds. Comp Biochem Physiol C Pharmacol Toxicol 112(1):1–14

Barter M, Cao L, Chen L, Lei G (2005) Results of a survey for waterbirds in the lower Yangtze floodplain, China, in January–February 2004. Forktail 21:1

Bird D, Tucker P, Fox GA, Lague P (1983) Synergistic effects of Aroclor® 1254 and mirex on the semen characteristics of American kestrels. Arch Environ Contam Toxicol 12(6):633–639

Blankenship AL, Kay DP, Zwiernik MJ, Holem RR, Newsted JL, Hecker M, Giesy JP (2008) Toxicity reference values for mink exposed to 2,3,7,8-tetrachlodibenzo-p-dioxin (TCDD) equivalents (TEQs). Ecotoxicol Environ Saf 69(3):325–349

Boily M, Champoux L, Bourbonnais D, Granges J, Rodrigue J, Spear P (1994) β-carotene and retinoids in eggs of Great Blue Herons (Ardea herodias) in relation to St Lawrence River contamination. Ecotoxicology 3(4):271–286

Bosveld A, Van den Berg M (1994) Effects of polychlorinated biphenyls, dibenzo-p-dioxins, and dibenzofurans on fish-eating birds. Environ Rev 2(2):147–166

Bosveld A, Gradener J, van Den Berg M, Murk A, Brouwer A, van Kampen M, Evers E (1995) Effects of PCDDs, PCDFs and PCBs in common tern (Sterna hirundo) breeding in estuarine and coastal colonies in the Netherlands and Belgium. Environ Toxicol Chem 14(1):99–115

Bosveld ATC, Nieboer R, de Bont A, Mennen J, Murk AJ, Feyk LA, Giesy JP, van den Berg M (2000) Biochemical and developmental effects of dietary exposure to polychlorinated biphenyls 126 and 153 in common tern chicks (Sterna hirundo). Environ Toxicol Chem 19(3): 719–730

Britton W, Huston T (1973) Influence of polychlorinated biphenyls in the laying hen. Poult Sci 52(4):1620–1624

Brunström B (1989) Toxicity of coplanar polychlorinated biphenyls in avian embryos. Chemosphere 19(1):765–768

Caldwell DJ, Mastrocco F, Hutchinson TH, Länge R, Heijerick D, Janssen C, Anderson PD, Sumpter JP (2008) Derivation of an aquatic predicted no-effect concentration for the synthetic hormone, 17α-ethinyl estradiol. Environ Sci Technol 42(19):7046–7054

CCME (1998) Protocol for derivation of Canadian tissue residue guidelines for the protection of wildlife consumers of aquatic biota. Canadian Council of Ministers of the Environment, Winnipeg

CCME (2001) Canadian tissue residue guidelines for the protection of wildlife consumers of aquatic biota: polychlorinated biphenyls(PCBs). Canadian Council of Ministers of the Environment, Winnipeg

Chang ES, Stokstad E (1975) Effect of chlorinated hydrocarbons on shell gland carbonic anhydrase and egg shell thickness in Japanese quail. Poult Sci 54(1):3–10

Coefield SJ, Fredricks TB, Seston RM, Nadeau MW, Tazelaar DL, Kay DP, Newsted J, Giesy JP, Zwiernik MJ (2010) Ecological risk assessment of great horned owls (Bubo virginianus) exposed to PCDD/DF in the Tittabawassee River floodplain in Midland, Michigan, USA. Environ Toxicol Chem 29(10):2341–2349

Connell D, Fung C, Minh T, Tanabe S, Lam P, Wong B, Lam M, Wong L, Wu R, Richardson B (2003) Risk to breeding success of fish-eating Ardeids due to persistent organic contaminants in Hong Kong: evidence from organochlorine compounds in eggs. Water Res 37(2):459–467

Custer TW, Heinz GH (1980) Reproductive success and nest attentiveness of mallard ducks fed Aroclor 1254. Environ Pollut A Ecol Biol 21(4):313–318

Eisler R (1986) Polychlorinated biphenyl hazards to fish, wildlife, and invertebrates: a synoptic review. U.S. Fish and Wildlife Service Biological Report 85(1.7). Contaminant Hazard Reviews

Elliott John E, Hart LE, Cheng KM, Norstrom RJ, Lorenzen A, Kennedy SW, Philibert H, Stegeman JJ, Bellward GD (1996) Biological effects of polychlorinated dibenzo-p-dioxins, dibenzofurans, and biphenyls in bald eagle (Haliaeetus leucocephalus) chicks. Environ Toxicol Chem 15(5):782–793

Elliott JE, Wilson LK, Henny CJ, Trudeau SF, Leighton FA, Kennedy SW, Cheng KM (2001) Assessment of biological effects of chlorinated hydrocarbons in osprey chicks. Environ Toxicol Chem 20(4):866–879

Fernie K, Smits J, Bortolotti G, Bird D (2001a) In ovo exposure to polychlorinated biphenyls: reproductive effects on second-generation American kestrels. Arch Environ Contam Toxicol 40(4):544–550

Fernie KJ, Smits JE, Bortolotti GR, Bird DM (2001b) Reproduction success of American kestrels exposed to dietary polychlorinated biphenyls. Environ Toxicol Chem 20(4):776–781

Fisher SA, Bortolotti GR, Fernie KJ, Bird DM, Smits JE (2006) Behavioral variation and its consequences during incubation for American kestrels exposed to polychlorinated biphenyls. Ecotoxicol Environ Saf 63(2):226–235

Fox LL, Grasman KA (1999) Effects of PCB 126 on primary immune organ development in chicken embryos. J Toxicol Environ Health A 58(4):233–244

Fredricks TB, Zwiernik MJ, Seston RM, Coefield SJ, Plautz SC, Tazelaar DL, Shotwell MS, Bradley PW, Kay DP, Giesy JP (2010) Passerine exposure to primarily PCDFs and PCDDs in the river floodplains near Midland, Michigan, USA. Arch Environ Contam Toxicol 58(4):1048–1064

Giesy JP, Ludwig JP, Tillitt DE (1994a) Deformities in birds of the Great Lakes region. Assigning causality. Environ Sci Technol 28(3):128–135

Giesy JP, Ludwig JP, Tillitt DE (1994b) Dioxins, dibenzofurans, PCBs and colonial, fish-eating water birds. In: Schecter A (ed) Dioxins and health. Plenum, New York, pp 249–307

Goff KF, Hull BE, Grasman KA (2005) Effects of PCB 126 on primary immune organs and thymocyte apoptosis in chicken embryos. J Toxicol Environ Health A 68(6):485–500

Gould JC, Cooper KR, Scanes CG (1999) Effects of polychlorinated biphenyls on thyroid hormones and liver type I monodeiodinase in the chick embryo. Ecotoxicol Environ Saf 43(2):195–203

Guruge K, Tanabe S, Fukuda M (2000) Toxic assessment of PCBs by the 2,3,7,8-tetrachlorodibenzo-p-dioxin equivalent in common cormorant (Phalacrocorax carbo) from Japan. Arch Environ Contam Toxicol 38(4):509–521

Hall LW, Scott MC, Killen WD (2009) Ecological risk assessment of copper and cadmium in surface waters of Chesapeake Bay watershed. Environ Toxicol Chem 17(6):1172–1189

Haseltine SD, Prouty RM (1980) Aroclor 1242 and reproductive success of adult mallards (*Anas platyrhynchos*). Environ Res 23(1):29–34

Heinz GH, Hill EF, Contrera JF (1980) Dopamine and norepinephrine depletion in ring doves fed DDE, dieldrin, and Aroclor 1254. Toxicol Appl Pharmacol 53(1):75–82

Hoffman DJ, Smith GJ, Rattner BA (1993) Biomarkers of contaminant exposure in common terns and black-crowned night herons in the Great Lakes. Environ Toxicol Chem 12(6):1095–1103

Hoffman DJ, Melancon MJ, Klein PN, Rice CP, Eisemann JD, Hines RK, Spann JW, Pendleton GW (1996) Developmental toxicity of PCB 126 (3,3′,4,4′,5-pentachlorobiphenyl) in nestling American kestrels (Falco sparverius). Toxicol Sci 34(2):188–200

Hoffman DJ, Melancon MJ, Klein PN, Eisemann JD, Spann JW (1998) Comparative developmental toxicity of planar polychlorinated biphenyl congeners in chickens, American kestrels, and common terns. Environ Toxicol Chem 17(4):747–757

Hong C-S, Bush B (1990) Determination of mono-and non-ortho coplanar PCBs in fish. Chemosphere 21(1):173–181

Kannan K, Blankenship A, Jones P, Giesy J (2000) Toxicity reference values for the toxic effects of polychlorinated biphenyls to aquatic mammals. Hum Ecol Risk Assess 6(1):181–201

Kennedy S, Lorenzen A, Jones S, Hahn M, Stegeman J (1996) Cytochrome P4501A induction in avian hepatocyte cultures: a promising approach for predicting the sensitivity of avian species to toxic effects of halogenated aromatic hydrocarbons. Toxicol Appl Pharmacol 141(1):214–230

Knight R, Randolph P, Allen G, Young L, Wigen R (1990) Diets of nesting bald eagles, Haliaeetus leucocephalus in western Washington. Can Field Nat 104(4):545–551

Kubiak TJ, Harris H, Smith L, Schwartz T, Stalling D, Trick J, Sileo L, Docherty D, Erdman T (1989) Microcontaminants and reproductive impairment of the Forster's tern on Green Bay, Lake Michigan-1983. Arch Environ Contam Toxicol 18(5):706–727

Lam JCW, Murphy MB, Wang Y, Tanabe S, Giesy JP, Lam PKS (2008) Risk assessment of organohalogenated compounds in water bird eggs from South China. Environ Sci Technol 42(16): 6296–6302

Lavoie E, Grasman K (2007) Effects of in ovo exposure to PCBs 126 and 77 on mortality, deformities and post-hatch immune function in chickens. J Toxicol Environ Health A 70(6):547–558

Leo Posthuma GW, Theo PT, Posthuma L, Suter GW, Traas TP (2002) Species sensitivity distributions in ecotoxicology, Environmental and ecological risk assessment. CRC Press, New York

Levengood JM, Schaeffer DJ (2010) Comparison of PCB congener profiles in the embryos and principal prey of a breeding colony of black-crowned night-herons. J Great Lakes Res 36(3): 548–553

Levengood JM, Wiedenmann L, Custer TW, Schaeffer DJ, Matson CW, Melancon MJ, Hoffman DJ, Scott JW, Talbott JL, Bordson GO (2007) Contaminant exposure and biomarker response in embryos of black-crowned night-herons (Nycticorax nycticorax) nesting near Lake Calumet, Illinois. J Great Lakes Res 33(4):791–805

Lillie RJ, Cecil HC, Bitman J, Fries G (1974) Differences in response of caged white leghorn layers to various polychlorinated biphenyls (PCBs) in the diet. Poult Sci 53(2):726–732

Lillie RJ, Cecil HC, Bitman J, Fries GF, Verrett J (1975) Toxicity of certain polychlorinated and polybrominated biphenyls on reproductive efficiency of caged chickens. Poult Sci 54(5):1550–1555

Linger JL, Peakall DB (1970) Metabolic effects of polychlorinated biphenyls in the American kestrel. Nature 228(5273):783–784

Ludwig JP, Giesy JP, Summer CL, Bowerman W, Aulerich R, Bursian S, Auman HJ, Jones PD, Williams LL, Tillitt DE (1993) A comparison of water quality criteria for the Great Lakes based on human and wildlife health. J Great Lakes Res 19(4):789–807

McKernan MA, Rattner BA, Hale RC, Ottinger MA (2007) Egg incubation position affects toxicity of air cell administered polychlorinated biphenyl 126 (3,3′,4,4′,5-pentachlorobiphenyl) in chicken (Gallus gallus) embryos. Environ Toxicol Chem 26(12):2724–2727

Nagy K (2001) Food requirements of wild animals: predictive equations for free-living mammals, reptiles, and birds. Nutr Abstr Rev B 71:1R–12R

Nelson AL, Martin AC (1953) Gamebird weights. J Wildl Manage 17(1):36–42

Newsted JL, Jones PD, Coady K, Giesy JP (2005) Avian toxicity reference values for perfluorooctane sulfonate. Environ Sci Technol 39(23):9357–9362

Peakall DB, Peakall ML (1973) Effect of a polychlorinated biphenyl on the reproduction of artificially and naturally incubated dove eggs. J Appl Ecol 10:863–868

Peakall DB, Lincer JL, Bloom SE (1972) Embryonic mortality and chromosomal alterations caused by Aroclor 1254 in ring doves. Environ Health Perspect 1:103

Peterson RE, Theobald HM, Kimmel GL (1993) Developmental and reproductive toxicity of dioxins and related compounds: cross-species comparisons. CRC Crit Rev Toxicol 23(3):283–335

Platonow N, Reinhart B (1973) The effects of polychlorinated biphenyls (Aroclor 1254) on chicken egg production, fertility and hatchability. Can J Comp Med 37(4):341–346

Powell D, Aulerich R, Meadows J, Tillitt D, Giesy J, Stromborg K, Bursian S (1996) Effects of 3,3′,4,4′,5-pentachlorobiphenyl (PCB 126) and 2,3,7,8-tetrachlorodibenzo-p-dioxin (TCDD) injected into the yolks of chicken (Gallus domesticus) eggs prior to incubation. Arch Environ Contam Toxicol 31(3):404–409

Powell DC, Aulerich RJ, Meadows JC, Tillitt DE, Powell JF, Restum JC, Stromborg KL, Giesy JP, Bursian SJ (1997) Effects of 3,3′,4,4′,5-pentachlorobiphenyl (PCB 126), 2,3,7,8-tetrachlorodibenzo-p-dioxin (TCDD), or an extract derived from field-collected cormorant eggs injected into double-crested cormorant (Phalacrocorax auritus) eggs. Environ Toxicol Chem 16(7):1450–1455

Powell DC, Aulerich RJ, Meadows JC, Tillitt DE, Kelly ME, Stromborg KL, Melancon MJ, Fitzgerald SD, Bursian SJ (1998) Effects of 3,3′,4,4′,5-pentachlorobiphenyl and 2,3,7,8-tetrachlorodibenzo-p-dioxin injected into the yolks of double-crested cormorant (Phalacrocorax auritus) eggs prior to incubation. Environ Toxicol Chem 17(10):2035–2040

Ricklefs RE (1979) Patterns of growth in birds. V. A comparative study of development in the starling, common tern, and Japanese quail. Auk 96:10–30

Roberts J, Rodgers D, Bailey J, Rorke M (1978) Polychlorinated biphenyls: biological criteria for an assessment of their effects on environmental quality. National Research Council Canada, Associate Committee on Scientific Criteria for Environmental Quality, Ottawa, ON

Safe S (1990) Polychlorinated biphenyls (PCBs), dibenzo-p-dioxins (PCDDs), dibenzofurans (PCDFs), and related compounds: environmental and mechanistic considerations which support the development of toxic equivalency factors (TEFs). CRC Crit Rev Toxicol 21(1):51–88

Safe SH (1994) Polychlorinated biphenyls (PCBs): environmental impact, biochemical and toxic responses, and implications for risk assessment. CRC Crit Rev Toxicol 24(2):87–149

Sample B, Opresko DM, Suter GW (1993) Toxicological benchmarks for wildlife. ORNL Oak Ridge National Laboratory, Oak Ridge, TN

Sanderson JT, Elliott JE, Norstrom RJ, Whitehead PE, Hart LE, Cheng KM, Bellward GD (1994) Monitoring biological effects of polychlorinated dibenzo-p-dioxins, dibenzofurans, and biphenyls in great blue heron chicks (Ardea herodias) in British Columbia. J Toxicol Environ Health A 41(4):435–450

Scott M (1977) Effects of PCBs, DDT, and mercury compounds in chickens and Japanese quail. In: Federation proceedings, p 1888

Senthilkumar K, Iseki N, Hayama S, Nakanishi J, Masunaga S (2002) Polychlorinated dibenzo-p-dioxins, dibenzofurans, and dioxin-like polychlorinated biphenyls in livers of birds from Japan. Arch Environ Contam Toxicol 42(2):244–255

Seston RM, Fredricks TB, Tazelaar DL, Coefield SJ, Bradley PW, Newsted JL, Kay DP, Fitzgerald SD, Giesy JP, Zwiernik MJ (2010) Tissue-based risk assessment of great blue heron (Ardea herodias) exposed to PCDD/DF in the Tittabawassee River floodplain, Michigan, USA. Environ Toxicol Chem 29(11):2544–2558

Strause KD, Zwiernik MJ, Im SH, Bradley PW, Moseley PP, Kay DP, Park CS, Jones PD, Blankenship AL, Newsted JL (2007) Risk assessment of great horned owls (Bubo virginianus) exposed to polychlorinated biphenyls and DDT along the Kalamazoo River, Michigan, USA. Environ Toxicol Chem 26(7):1386–1398

Struger J, Weseloh DV (1985) Great Lakes Caspian terns: egg contaminants and biological implications. Colonial Waterbirds 8:142–149

Tillitt DE, Ankley GT, Giesy JP, Ludwig JP, Kurita-Matsuba H, Weseloh DV, Ross PS, Bishop CA, Sileo L, Stromborg KL (1992) Polychlorinated biphenyl residues and egg mortality in double-crested cormorants from the great lakes. Environ Toxicol Chem 11(9):1281–1288

Tori GM, Peterle TJ (1983) Effects of PCBs on mourning dove courtship behavior. Bull Environ Contam Toxicol 30(1):44–49

Tremblay J, Ellison LN (1980) Breeding success of the black-crowned night heron in the St. Lawrence Estuary. Can J Zool 58(7):1259–1263

Tumasonis CF, Bush B, Baker FD (1973) PCB levels in egg yolks associated with embryonic mortality and deformity of hatched chicks. Arch Environ Contam Toxicol 1(4):312–324

USEPA (1985) Guidelines for deriving numerical national water quality criteria for the protection of aquatic organisms and their uses. United States Environmental Protection Agency, Office of Research and Development, Duluth, MN

USEPA (1995a) Great lakes water quality initiative criteria documents for the protection of wildlife. EPA-820-B-95-008. Office of Water, Washington, DC

USEPA (1995b) Great lakes water quality initiative technical support document for wildlife criteria. EPA-820-B-95-009. Office of Water, Washington, DC

USEPA (2005) Science Advisory Board Consultation Document. Proposed revisions to aquatic life guidelines: tissue-based criteria for "bioaccumulative" chemicals. Office of Water, Washington, DC

USEPA (2007) Ecological soil screening level for DDT and metabolites. Office of Solid Waste and Emergency Response, Washington, DC

Van den Berg M, Birnbaum L, Bosveld A, Brunström B, Cook P, Feeley M, Giesy JP, Hanberg A, Hasegawa R, Kennedy SW (1998) Toxic equivalency factors (TEFs) for PCBs, PCDDs, PCDFs for humans and wildlife. Environ Health Perspect 106(12):775

Wei X, Leung K, Wong M, Giesy J, Cai Z, Wong CKC (2011) Assessment of risk of PCDD/Fs and dioxin-like PCBs in marine and freshwater fish in Pearl River Delta, China. Mar Pollut Bull 63(5–12):166–171

Weseloh D, Hamr P, Bishop CA, Norstrom RJ (1995) Organochlorine contaminant levels in waterbird species from Hamilton Harbour, Lake Ontario: an IJC area of concern. J Great Lakes Res 21(1):121–137

Wiemeyer SN, Lamont TG, Bunck CM, Sindelar CR, Gramlich FJ, Fraser JD, Byrd MA (1984) Organochlorine pesticide, polychlorobiphenyl, and mercury residues in bald eagle eggs—1969–79—and their relationships to shell thinning and reproduction. Arch Environ Contam Toxicol 13(5):529–549

Wiemeyer SN, Bunck CM, Stafford CJ (1993) Environmental contaminants in bald eagle eggs—1980–84—and further interpretations of relationships to productivity and shell thickness. Arch Environ Contam Toxicol 24(2):213–227

Wiesmüller T, Schlatterer B, Wuntke B, Schneider R (1999) PCDDs/PCDFs, coplanar PCBs and PCBs in barn owl eggs from different areas in the state of Brandenburg, Germany. Bull Environ Contam Toxicol 63(1):15–24

Yamashita N, Tanabe S, Ludwig JP, Kurita H, Ludwig ME, Tatsukawa R (1993) Embryonic abnormalities and organochlorine contamination in double-crested cormorants (*Phalacrocorax auritus*) and Caspian terns (*Hydroprogne caspia*) from the upper Great Lakes in 1988. Environ Pollut 79(2):163–173

Zhang R, Wu F, Li H, Guo G, Feng C, Giesy JP, Chang H (2013) Toxicity reference values and tissue residue criteria for protecting avian wildlife exposed to methylmercury in China. Rev Environ Contam Toxicol 223:53–80

# Fabricated Nanoparticles: Current Status and Potential Phytotoxic Threats

Tushar Yadav, Alka A. Mungray, and Arvind K. Mungray

## Contents

| | | |
|---|---|---|
| 1 | Introduction | 84 |
| 2 | Entry of Nanoparticles into the Environment | 87 |
| 3 | Factors Affecting Nanoparticle Phytotoxicity | 88 |
| | 3.1 Environmental Factors | 89 |
| | 3.2 Stabilizers and Dispersion Medium | 89 |
| | 3.3 Particle Size | 90 |
| | 3.4 Surface Characteristics and Concentration | 90 |
| 4 | Nanoparticle Entry into Plant Systems | 91 |
| 5 | Phytotoxic Effects of Nanoparticles | 92 |
| | 5.1 Single-Walled Carbon Nanotubes (SWCNTs) and Multiwalled Carbon Nanotubes (MWCNTs) | 92 |
| | 5.2 Titanium Nanoparticles | 96 |
| | 5.3 Zinc Nanoparticles | 97 |
| | 5.4 Silver Nanoparticles | 98 |
| | 5.5 Copper Nanoparticles | 99 |
| | 5.6 Cerium Nanoparticles | 99 |
| | 5.7 Molybdenum Nanoparticles | 100 |
| | 5.8 Gold Nanoparticles | 101 |
| | 5.9 Magnetic Nanoparticles | 101 |
| | 5.10 Other Nanoparticles | 101 |
| 6 | Discussion | 102 |
| 7 | Summary | 104 |
| | References | 105 |

T. Yadav • A.A. Mungray • A.K. Mungray (✉)
Chemical Engineering Department, Sardar Vallabhbhai National Institute of Technology, Ichchhanath, Surat, Gujarat 395007, India
e-mail: akm@ched.svnit.ac.in

## 1 Introduction

Nanotechnology is a relatively new technology that involves manipulating matter on an atomic and molecular scale. In general, nanotechnology deals with materials, devices, and other structures having at least one dimension in a size range from 1 to 100 nm (Roco 2003; SCENIHR 2005; Moore 2006). The recent growth in this sector has promised several benefits to society by exploiting the novel properties of nanoparticles. Nanotechnology offers an array of potential applications, and is becoming a key technology for the upcoming generation. Billions of dollars have been invested in nanotechnology research and development across the world. For instance, in the USA, the National Nanotechnology Initiative has invested $3.7 billion, whereas, respectively, the European Union (EU) and Japan have respectively invested $1.2 billion and $750 million dollars in this technology (ANUI 2012). Today, nanotechnology is increasingly occupying a prominent position in human life and in human lifestyle. Moreover, the development of nanomaterials and nanodevices has opened many novel applications in science and technology.

Nanomaterials may be defined as containing constituent particles that have nanoscale dimensions, and are fabricated to remain as surface-bounded, dispersed state or aggregate forms. Nanoparticles (NPs) can be categorized on the basis of their origin, viz., natural, incidental, and fabricated (Bhatt and Tripathi 2011; Farre et al. 2011; Smita et al. 2012) as described in Table 1. Nanoparticles have existed in nature from the beginning of the earth's history and they are still found in the environment in the form of volcanic dust, lunar dust, mineral composites, etc. (Rietmeijer and Mackinnon 1997; Reid et al. 2000; Verma et al. 2002; Lee and Richards 2004).

**Table 1** A list of naturally and anthropogenically produced nanoparticles

| Nanoparticles | Particle type | References |
|---|---|---|
| Natural | | |
| Volcanic dust | Bismuth oxide | Rietmeijer and Mackinnon (1997) |
| Volcanic ash | Crystalline silica | Lee and Richards (2004) |
| Ocean surface microlayer | Colloids, carbon particles | El Nemr and Abd-Allah (2003), Wigginton et al. (2007) |
| Soil | Mineral particles, colloids | Reid et al. (2000) |
| Ice cores | Carbon nanotubes, fullerenes, silicon dioxide | Murr et al. (2004) |
| Historic sediments | Hematite, silicates | Verma et al. (2002) |
| Forest fire | Carbon particles | Smita et al. (2012) |
| Incidental | | |
| Vehicle exhaust, coal/oil/gas boiler, wielding | Carbon particles, colloids | Novack and Bucheli (2007), Remedios et al. (2012), Smita et al. (2012) |
| Fabricated | | |
| Drug delivery, diagnostic, ground water remediation, industrial production process | Silicon dioxide, silver, zero valent iron, cerium oxide, etc. | Farre et al. (2011), Remedios et al. (2012), Smita et al. (2012) |

Incidental nanoparticles, also defined as anthropogenic nanoparticle waste, are produced as a result of human activities such as industrial processes, coal combustion, welding fumes, vehicle exhaust, etc. (Novack and Bucheli 2007; Smita et al. 2012). In contrast, fabricated nanoparticles are designed and produced to achieve specific physicochemical properties targeted towards unique applications. Fabricated nanoparticles comprise four major types:

1. Carbon-based nanomaterials that usually include fullerene, single walled carbon nanotubes (SWCNT) and multiwalled carbon nanotubes (MWCNT), or nanowires
2. Metal-based nanomaterials such as quantum dots, nanogold, nanosilver, nanoiron, and nanoscale metal oxides like $TiO_2$, $ZnO$, and $CeO_2$
3. Dendrimers that are nano-sized polymeric structures constructed from branched units that are capable of being customized to achieve explicit biological and chemical functions (Klajnert and Bryszewska 2001)
4. Nanocomposites that are combinations of different nanoparticles or nanoparticles with larger bulk-type materials (Lin and Xing 2007), and are fabricated to have different morphologies such as spheres, rods, tubes and prisms (Yu-Nam and Lead 2008).

Fabricated nanoparticles have received intense interest as a result of their positive impact and wide applicability in several sectors of the economy (Table 2). Examples of fabricated NP industrial applications include the following: electronics, textiles, pharmaceuticals, cosmetics, water treatment technology, and energy and agriculture, among others. Progressively, more nanochemicals are being produced and are being slated for use in various new industrial applications (Novack and Bucheli 2007). A Swiss survey (Schmid and Riediker 2008) indicated that the estimated annual production of nanomaterials was 2,419 t. This figure is much higher than estimates made by the Royal Society and those in a Royal Academy of Engineering report on Nanotechnology (2004) for the year 2010. This suggests that the actual worldwide production of nanomaterials may be higher than thought. The reason is that existing production quantities of nanomaterials are not well known, and the production rates of these nanomaterials and associated products have rapidly increased from kilograms to thousands of tons in a relatively few decades. Of course, this rapid introduction of new nanomaterials will result in increased future environmental discharge of these materials.

Because of the volumes being produced, concern is growing about the environmental risks of many fabricated nanomaterials (USEPA 2007). Work is proceeding to evaluate the potential undesirable risks of nanomaterials on the environment and on human well-being. The concern for nanotechnology-derived risks has given rise to a new field of study called nanotoxicity. Nanotoxicity addresses the toxicity of nanomaterials to various life forms and to the environment.

One may legitimately ask why NPs are of such concern, since certain forms of them already exist naturally in the environment (e.g., volcanic dust, forest fire, soil; Rietmeijer and Mackinnon 1997; Reid et al. 2000; Lee and Richards 2004; Smita et al. 2012). In response, it is clear that natural substances may also be hazardous. Indeed, naturally occurring nanoparticles under certain circumstances may harm

**Table 2** An inventory of various fabricated nanoparticles and their commercial applications

| Fabricated NPs | Field of applications | References |
| --- | --- | --- |
| Carbon nanotubes and derivatives | Electronics and computers, catalyst, battery, fuel cell electrodes, supercapacitors, etc. | Pulickel and Zhou (2001), Bhatt and Tripathi (2011) |
| $TiO_2$ | Cosmetics, skin care products, sunscreen lotions, solar cells, paints and coatings | Bhatt and Tripathi (2011), Gupta and Tripathi (2011) |
| ZnO | Skin care products, bottle coatings, etc. | Bhatt and Tripathi (2011), Smijs and Pavel (2011) |
| $CeO_2$ | Combustion catalyst in diesel fuels to improve emission quality, gas sensor, solar cells, oxygen pumps | Kosynkin et al. (2000), Corma et al. (2004), Livingston and Helvajian (2005), Bhatt and Tripathi (2011) |
| Silver NPs | Disinfectants, wound dressings, antibacterial clothing and spray | Bosetti et al. (2002), Yeo et al. (2003), Cho et al. (2005), Bhatt and Tripathi (2011) |
| Gold NPs | Cancer therapy, sensors, catalyst, flexible conducting inks and films | Peng et al. (2010), Perrault and Chan (2010), Bhatt and Tripathi (2011) |
| Iron and iron oxide | In medical imaging, cleaning up groundwater pollution | Huber (2005), Bhatt and Tripathi (2011) |
| Silica | Thermal and electrical insulators, drug delivery, etc. | Foraker et al. (2003), O'Farrell et al. (2006) |
| Quantum dots | Medicine (medical imaging and targeted therapeutics), in solar cells, photovoltaic cells, security inks, photonics, and telecommunications | Howarth et al. (2008), Konstantatos and Sargent (2009), Bhatt and Tripathi (2011), Hoshino et al. (2012) |
| Dendrimers | Manufacture of macrocapsules, nanolatex, colored glasses, drug delivery and DNA chips, tumor treatment | Roy et al. (1993), Capala et al. (1996), Twyman et al. (1999), Kukowska-Latallo et al. (2000), Bhatt and Tripathi (2011) |

life forms. In addition, the production and use on a large scale of man-made nanomaterials will increase their environmental concentrations and human exposure to them. Such increased environmental concentrations will concomitantly increase the possibility of harmful interactions between nanoparticles and life forms. Some research reports also exist on the interaction with, and adsorption of environmental pollutants by nanomaterials (Cheng et al. 2004; Yang et al. 2006; Gotovac et al. 2007; Hu et al. 2008). Owing to their minute size and high surface reactivity, NPs may traverse cell barriers of living organisms and interact with intracellular entities. Therefore, they may contribute to potential cellular and genetic turmoil. Yet greater volumes are incrementally being commercially manufactured and released into the open environment, irrespective of health concerns or the need for prior safety assessment for environmental or health impact. To properly deal with the safety of nanotechnology in the future, we will require a multidisciplinary approach that must enjoin and enlist scientists, risk assessors, regulators, and policy makers in scientific debate and action.

Plants are a critical base life form of all ecosystems and have a significant position in trophic transfer and maintenance of worldwide ecological balance. Environmental conditions highly influence plant growth and viability. Therefore, exposure of plants to certain natural or xenobiotic substances above a certain optimum concentration may cause toxicity. In addition, toxic substances that have no known function in plant systems are nonetheless accumulated in plant tissues, with potential lethal effects for non-tolerant species. Plants have evolved in the presence of several natural nanomaterials (Rietmeijer and Mackinnon 1997; Reid et al. 2000; Lee and Richards 2004; Smita et al. 2012). However, as the production and use of fabricated nanomaterials has increased, the probability of plant exposure to nanomaterials (NMs) has greatly increased (Ruffini and Cremonini 2009; Rico et al. 2011).

The uptake, accumulation, translocation, and toxicity of nanoparticles in plant systems are a very recently formed field of research. Researchers have reported positive, negative, and/or inconsequential effects from plants being exposed to nanoparticles. NP-associated alteration of morphological features such as effects on roots, leaves and seed germination have been reported. Unfortunately, to date, very few studies have been conducted on the genetic response of plants that are exposed to nanoparticles.

In this review, we summarize recent findings on the potential phytotoxic threats posed by fabricated NPs. In addition, we discuss the various factors that affect the phytotoxicity of nanoparticles and recommend how the challenges presented by nanoparticle toxicity to plants can be addressed.

## 2 Entry of Nanoparticles into the Environment

As mentioned earlier nanoparticles exist in the environment naturally; however, natural nanoparticles are present at very low concentrations and have negligible impact (Klaine et al. 2008; Remedios et al. 2012). In recent decades, fabricated NPs have emerged and have been incorporated into a growing number of commercial products. The effort, both scientific and commercial, to better understand the threat posed by nanoparticles and to control their discharge into the environment has become very large. Although, in many cases, the release of nanoparticles is unavoidable, the goal should be to minimize those releases of NMs that could pose a significant risk to the environment or to humans.

In Fig. 1, we present an outline of the major possible pathways through which fabricated nanoparticles may enter into environment to potentially cause toxicity to plants.

Nanowastes are released from many different human activities, such as research and development, industrial production, transport and storage, and primarily after disposal of consumer products that contain nanomaterials (Novack and Bucheli 2007; Farre et al. 2011; Smita et al. 2012). Nanowastes may enter the environment directly or via municipal or industrial waste treatment plants (WTPs). Currently WTPs are not efficient in removing nanoparticles from waste streams or from water. Once released to the environment, nanowastes accumulate in ecosystems and pose

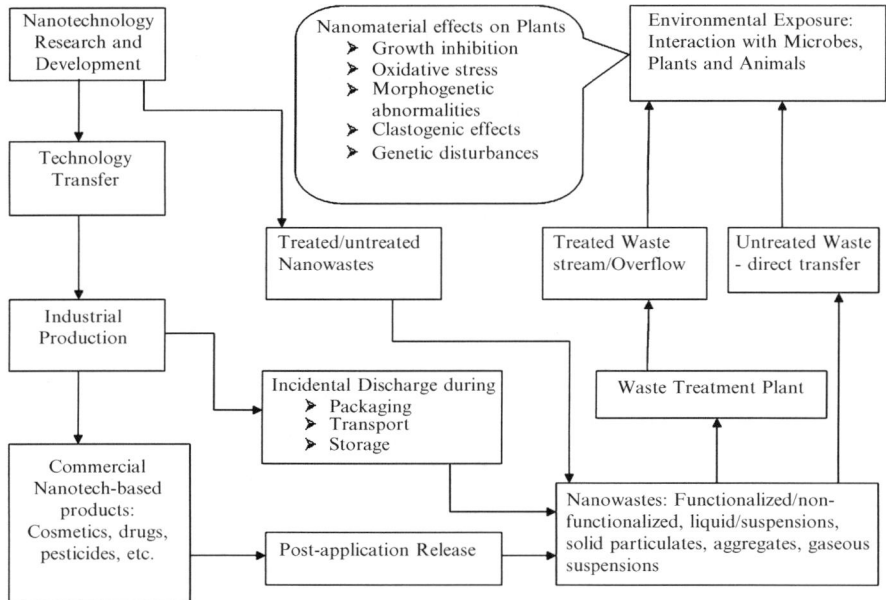

**Fig. 1** An outline of how nanowaste flows in the environment to potentially produce phytotoxic effects

threats to life forms. Incidental discharge of NPs to the atmosphere occurs from combustion processes, from boilers and power plants, and from vehicle exhaust releases. In some cases, NPs are directly released to the environment to achieve environmental cleanliness, an example of which is the use of $TiO_2$ NPs in pilot water-purification reactors (Shahmoradi et al. 2010; Larue et al. 2011).

Many new studies on nanoparticle safely have been initiated, and many others completed that address prospective nanotoxic effects on human health. However, few studies are yet available on the potential ecotoxic effects of NPs. More ecotoxicity data are certainly needed, because NPs may destabilize or disrupt ecosystems by entering food chains via plant ingestion and trophic transfer. Unfortunately, data on NP accumulation and toxicity in plants are still limited. To date, studies that have addressed effects of nanotoxicants to plants have emphasized impairment of seed germination and root elongation (Lin and Xing 2007; Lee et al. 2008; Oleszczuk et al. 2011; Ruffini et al. 2011; Wang et al. 2012; Wu et al. 2012)

## 3 Factors Affecting Nanoparticle Phytotoxicity

Our review of the literature revealed that the uptake, translocation, and accumulation of NPs depends on the species of plant, and the size, type, chemical composition, functionalization, and stability of the NPs in a system (Buzea et al. 2007; Lin and Xing 2007; Lee et al. 2008; Barrena et al. 2009; Larue et al. 2012).

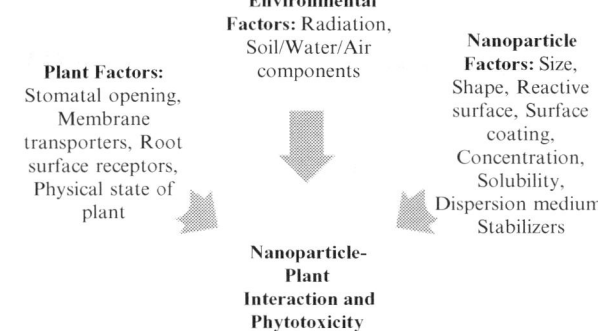

**Fig. 2** Various factors mediating plant–nanoparticle interactions that produce phytotoxicity

Fabricated NPs are synthesized for specific purposes; hence, the physicochemical properties of each NP vary significantly. Features like size, shape, and surface characteristics may augment or change the reactivity of a previously inert bulk material, and in turn, this change may generate a toxic response (Kumar et al. 2012). Changed features may also affect the likelihood of trespassing natural barriers, and may affect solubility and mobility in liquid media or in air, soil, etc. (Buzea et al. 2007; Somasundaran et al. 2010; Aubert et al. 2012).

Once released into the open environment, nanoparticles may chemically react in various ways, or may undergo photo-induced chemical changes. For metal and metal-oxide nanoparticles such processes can transform the particle to render it more hygroscopic and potentially more soluble as metal ions. Such transformed NPs may have diverse and unusual physicochemical properties, and they may be deposited onto aqueous and terrestrial ecosystems. Understanding the details of such transformations is key to understanding how NPs behave in the environment (Rubasinghe et al. 2010). Above, in Fig. 2, we present an outline of the factors that influence the interactions between plants and nanoparticles that may lead to phytotoxicity.

## 3.1 *Environmental Factors*

Nanoparticles exposed to the environment may be transformed by moisture, sunlight, soil components, and by the action of living organisms, among other factors. Some fabricated nanoparticles (e.g., $TiO_2$) show photocatalytic activity on exposure to UV light and may generate Reactive Oxygen Species (ROS) that produce genetic alterations (i.e., breaks in the nucleic acid chain or cross-linking to form adducts with bases or sugars; Cabiscol et al. 2000; Khus et al. 2006; Zhao et al. 2007).

## 3.2 *Stabilizers and Dispersion Medium*

Stabilizers are coatings used to preserve specific properties that NPs display. Barrena et al. (2009) reported that metallic NPs had low to zero toxic effects on two vegetables, lettuce and cucumber. They suggested that the presence of stabilizers significantly affected the behavior of NPs. García et al. (2011) studied the effect of stabilizers on NP toxicity by taking germination index (rate and extent of seed germination) as a key measure and found reduced germination. Similarly the response of bare- and alginate-coated $CeO_2$ NPs in soil and their uptake by corn plants was investigated by Zhao et al. (2012). The authors of this study concluded that the uptake of coated $CeO_2$ NPs was increased in treated plants vs. non-coated NPs. In another study, Oleszczuk et al. (2011) demonstrated that the dispersion medium may modulate NP toxicity. They investigated the phytotoxicity of various sewage sludges containing MWCNTs and found that the type of sludge determines the level of toxicity. In natural ecosystems the uptake rates and toxicity of NMs by plants are expected to be dependent on the chemical properties, organic content, and the colloidal properties of the associated dispersion medium, such as soil, sludge, or sediments.

## 3.3  Particle Size

Larue et al. (2012) suggested that smaller NPs may accumulate in wheat roots, and without dissolution or crystal phase modification, may disperse throughout the tissues of the treated plant. Barrena et al. (2009), concluded, after reviewing many research articles, that particle size and specific surface area were the most suitable factors for evaluating NP phytotoxicity. Small-sized NPs may be toxic or may become more toxic to plants (Ma et al. 2010; Lopez-Moreno et al. 2010b; Shen et al. 2010; Vochita et al. 2012). Such small particles are naturally taken up by plants via sieving and then pass into the protoplasm (Navarro et al. 2008a). NP interaction at the plant surface appears to enhance formation of large new pores that further facilitates nanoparticle uptake by plants (Navarro et al. 2008a, b). One possible pathway reported for silver NPs (<20 nm) was transport through the plasmodesmata to the cell interior (Ma et al. 2010).

## 3.4  Surface Characteristics and Concentration

The surface characteristics of NPs may play a significant role in producing phytotoxicity. Surface characteristics affect aggregation properties and mobility in aquatic and terrestrial systems, and therefore affect interactions with plants (Klaine et al. 2008; Navarro et al. 2008a, b; Auffan et al. 2010). Yang and Watts (2005) confirmed this by loading (adsorbing) phenanthrene onto the surface of alumina NPs, before testing their phytotoxicity. Lower plant toxicity was observed for loaded than for non-loaded particles. However, after uptake of these surface modified NPs, toxicity could be either result from the chemical compound itself that was adsorbed onto the NP or from the synergistic action of both (Novack and Bucheli 2007; Zhao et al. 2012).

Generally, for most NPs, relatively high concentrations are required to produce detectable plant toxicity, and the threshold of any toxicity that appears is species dependent (Lin and Xing 2007; Lee et al. 2008).

## 4 Nanoparticle Entry into Plant Systems

Several studies have been performed to investigate the entry of NPs into plant systems (Eichert et al. 2008; Somasundaran et al. 2010; Majumdar and Ahmed 2011; Moaveni et al. 2011), although the phenomenon is still poorly understood. It is known that NPs may form complexes with membrane transporter proteins or root exudates, and as a consequence, subsequently be transported into the plant system. Majumdar and Ahmed (2011) identified the involvement of functional groups such as carboxyl, hydroxyl, amine, carbonyl, etc., in binding silver NPs to root cells. The binding of nanoparticles in roots may result from complex formation with certain functional groups, from physical adsorption, chemical reactions with surface sites, or ion exchange and surface precipitation (Gupta and Rastogi 2008; Srividya and Mohanty 2009).

Jia et al. (2005) has stated that transport of NPs across membrane occurs via embedded transport carrier proteins or through ion channels. Nanoparticles that interact with the surface membrane and are taken up by cells normally follow this explicit biological mechanism (Nair et al. 2010; Moaveni et al. 2011). Moreover, the intrinsic characteristics of NPs such as roughness, hydrophobicity and charge lead to nonspecific binding forces that promote surface binding and cellular uptake. In contrast, if there is involvement of a specific receptor–ligand interaction, absorption into the cell may occur by an endocytic uptake process. NPs that have a spiky surface may penetrate the cell membrane directly, without involving an endocytic pathway as a result of the combined effect of nonspecific binding forces on their surfaces (Somasundaran et al. 2010). Inside the cell, these NPs may bind with various cytoplasmic organelles and interfere with metabolic processes (Jia et al. 2005). Their entry and interaction with subcellular structures produce oxidative stress (Unfried et al. 2007).

Several reports (Eichert et al. 2008; Fernandez and Eichert 2009; Uzu et al. 2010) provide evidence that NPs that are accumulated on leaf surfaces may enter the plant through stomatal openings or via the bases of trichomes, after which they may disperse to various tissues. How NPs enter plants through the cell wall still remains unsolved. However, one phenomenon that may be critical to a successful penetration is the orientation of the NPs with respect to the plant cell wall, and this particular phenomenon needs further study (Moaveni et al. 2011). The entry of NPs into plants may also be affected by NP surface characteristics. Adsorption of toxic entities onto NP surfaces, surface coatings, and soil organic matter were reported to enhance their uptake into higher plants (Zhao et al. 2012).

Why some plant species readily take up several NPs and others do not is still unknown and must be further explored. We postulate that such differential accumulation may be explained by differences in root microstructures of different plants, and/or the physical and chemical interactions that occur between the NPs and the root exudates in the rhizosphere.

## 5 Phytotoxic Effects of Nanoparticles

In this section we summarize the phytotoxicological aspects of NPs and how they affect the plant system. In Table 3, we summarize the toxic effects that NPs produce on different plant species.

### 5.1 Single-Walled Carbon Nanotubes (SWCNTs) and Multiwalled Carbon Nanotubes (MWCNTs)

As a result of their small size, carbon nanotubes are hypothesized to interact with proteins and polysaccharides on the cell wall to elicit hypersensitive responses similar to those produced by plant pathogens. This eventually leads to cell mortality (Tan and Fugetsu 2007; Lin et al. 2009a, b; Tan et al. 2009). Lin et al. (2009a, b) examined the uptake and translocation of carbon NPs by rice plants (*Oryza sativa*). They found that fullerene $C_{70}$ might easily be taken up by roots and transported to shoots. This study also proved that, if $C_{70}$ entered plants through the leaves, it may possibly be transported downward from leaves to roots through the phloem. Lin et al. (2009a, b) suggested that the presence of metallic impurities (i.e., residual metal catalysts, etc.) during the synthesis of CNTs may also contribute to toxicity.

Shen et al. (2010) studied the impact of SWCNTs in rice and *Arabidopsis* protoplast cells. The results were that the nanoscale size and concentration of SWCNTs were major characteristics responsible for potential cytotoxic effects. Profuse endonucleolytic cleavage of DNA was evident in the *Arabidopsis* cells, indicating the genotoxic potential of SWCNTs in plant system.

Lin and Xing (2007) soaked plant seeds in a MWCNT suspension for 2 h and observed the response. They concluded that no significant phytotoxic effect or physiological response was notable after MWCNTs were applied. Instead, seed germination was accelerated, and percent germination rate and vegetative mass were observed to increase (Khodakovskaya et al. 2009). This response may have resulted from increased water uptake induced by contact with the CNTs. The possible toxic effects of MWCNTs on suspended rice cells (*Oryza sativa* L.) was investigated by Tan et al. (2009). They demonstrated decreased cell viability from increased accumulation of ROS. This response to MWCNTs exposure was dose-dependent and produced reduced cell density of the cultured rice cell suspension.

Canas et al. (2008) studied the phytotoxicity of nanotubes functionalized with poly-3-aminobenzenesulfonic acid. In some cases, they found that root lengths were affected more by non-functionalized carbon nanotubes than by functionalized

Table 3 A survey of various toxic effects known to be caused by NPs on different plant species

| Nanoparticles type | Average NP size | Concentration range | Plant group affected | Phytotoxic effects | References |
|---|---|---|---|---|---|
| SWCNT | 1–2 nm | – | Rice | Delayed flowering, decrease yield, cytotoxicity | Lin et al. (2009a, b), Shen et al. (2010) |
|  | 1–2 nm | 5–250 mg/L | Arabidopsis thaliana | DNA breakdown | Shen et al. (2010) |
| MWCNT |  | – | Zucchini | Reduced biomass | Stampoulis et al. (2009) |
|  | 10–20 nm | 2,000 mg/L | Lettuce | Reduced root length | Lin and Xing (2007) |
|  | 10–30 nm | 20 mg/L | Rice | Chromatin condensation and plasma membrane detachment from cell wall, cell shrinkage, and cell death | Tan et al. (2009) |
|  | 20–70 nm | 100–1,000 mg/L | Onobrychis arenaria | Mechanical injury, inhibition of peroxidases, oxidative stress | Smirnova et al. (2011) |
|  | 10–60 nm | 100–5,000 mg/L | Raphanus sativus, Cucumis sativus | Germination inhibition | Oleszczuk et al. (2011) |
| TiO$_2$ NPs | 20–100 nm | – | Wheat | Decrease in biomass, cell membrane damage | Heinlaan et al. (2008), Du et al. (2011) |
|  | 100 nm | 20,000–40,000 mg/L, 2,000 mg/L | Vicia narbonensis L., Zea mays L. | Delayed germination and root elongation, reduced MI, increased AI | Ruffini et al. (2011) |
|  | 100 nm | 319 mg/L | Allium cepa | DNA damage, growth inhibition, and increased lipid peroxidation | Ghosh et al. (2010) |
|  | 100 nm | 157 mg/L | Nicotiana tabacum | DNA damage | Ghosh et al. (2010) |
|  | <50 nm | 5–50 mg/L | Vicia faba | Root surface accumulation, blocking cell connection and cell wall pores, decrease in shoot biomass, decreased GR and APX activities in roots | Ovecka et al. (2005), Anne-Sophie et al. (2011) |
|  | 30 nm | 1,000 mg/L | Maize (Zea mays L.) | Reduction in root cell pore diameter, reduced transpiration and leaf growth | Asli and Neumann (2009) |
|  | – | – | Arabidopsis thaliana | Disruption of microtubular network | Wang et al. (2011) |
| ZnO NPs | 40–100 nm | – | Wheat | Decrease in biomass | Du et al. (2011) |

(continued)

Table 3 (continued)

| Nanoparticles type | Average NP size | Concentration range | Plant group affected | Phytotoxic effects | References |
|---|---|---|---|---|---|
| | 45 nm | 10–2,000 mg/L | Buckwheat (*Fagopyrum esculentum*) | Decrease in seedling biomass, root cell damage, uncontrolled induction of ROS defense system | Lee et al. (2013) |
| | 20–25 nm | 20–50 mg/L | Ryegrass, rapeseed, radish | Reduced biomass, shrank root tips, epidermis and rootcap were broken, highly vacuolated and collapsed cortical cells | Lin and Xing (2007), Lin and Xing (2008a) |
| | 8 nm | 4,000 mg/L | Soyabean (*Glycine max*) | DNA unstability, differential effect on plant growth and element uptake | Lopez-Moreno et al. (2010b) |
| | 20 nm | 10–2,000 mg/L | *Cicer arietinum*, *Vigna radiate* | Growth inhibition | Mahajan et al. (2011) |
| | 44–50 nm | 400–4,000 mg/L | *Arabidopsis thaliana* | Reduced seed germination, leaf number, and root elongation | Lee et al. (2010) |
| Zn NPs | 13 nm | 2,000 mg/L | Radish, rape, ryegrass, lettuce, corn, cucumber | Highly reduced root growth | Lin and Xing (2007) |
| Ag NPs | – | 100–1,000 mg/L | *Cucurbita pepo* | Reduced growth, transpiration, and biomass | Stampoulis et al. (2009) |
| | 25 nm | 100 mg/L | *Oryza sativa* | Cell wall damage, vacuole damage | Majumdar and Ahmed (2011) |
| | <100 nm | 25–100 mg/L | *Allium cepa* | Decreased MI, increased chromosomal abnormalities and aberrations | Babu et al. (2008), Kumari et al. (2009) |
| | 24 nm | 0.01–0.1 mg/L and 10–100 mg/L | *Bacopa monnieri* (Linn.) | Disappearance of air chamber in root cortex, alteration of shape, size, and distribution of xylem elements in stem | Krishnaraj et al. (2012) |
| | 60 nm | 12.5–100 mg/L | *Vicia faba* | Decrease in mean MI, increased chromatid breaks, isochromatic breaks, acentric fragments, micronuclei, etc. | Patlolla et al. (2012) |
| Cu NPs | – | 251–447 mg/L | Mung bean (*Phaseolus radiates*) | Reduced seedling growth rate | Lee et al. (2008) |
| | – | 450–722 mg/L | Wheat (*Triticum aestivum*) | Reduced seedling growth rate | Lee et al. (2008) |
| | – | 1,000 mg/L | Zucchini | Reduced root length and biomass | Stampoulis et al. (2009) |
| CuO NPs | 20–40 nm | 100 mg/L | *Zea mays* | Seedling growth inhibition | Wang et al. (2012) |

| Nanoparticle | Size | Concentration | Plant | Effect | Reference |
|---|---|---|---|---|---|
| CeO$_2$ NPs | 30–50 nm | – | Lettuce, radish, cucumber | Seed germination inhibition | Wu et al. (2012) |
| | – | 500–4,000 mg/L | Zea mays, Cucumis sativus, Lycopersicon esculentum | Reduced germination index | López-Moreno et al. (2010a) |
| | 7 nm | 2,000–4,000 mg/L | Glycine max | Alterations in DNA, differential effect on plant growth and element uptake | Lopez-Moreno et al. (2010b) |
| | 6.5 nm | 640 mg/L | Lactuca sativa, Cucumis sativus, Solanum lycopersicum, Spinacia oleracea | Reduced germination | García et al. (2011) |
| Mo NPs, H$_2$O-CMB and EtOH-CMB | 2.3 μm and 550 nm | 0.0051–0.51 mg/L | Rapeseed (Brassica napus) | Damages root morphology, loss of gravitropism, stunted plant growth | Aubert et al. (2012) |
| Au NPs | 3.5–18 nm | – | Nicotiana xanthi | Leaf necrosis | Sabo-Attwood et al. (2012) |
| Magnetic NPs, ZnFe$_2$O$_4$, CoFe$_2$O$_4$ and Fe$_3$O$_4$ | 11.4 nm, 7.5 nm and 9.7 nm respectively | 102–510 mg/L | Sunflower | Reduction in MI, increased AI | Vochita et al. (2012) |
| Fe$_3$O$_4$ NPs | 6 nm | 2.01–33.5 mg/L | Daucus carota L. | Affected growth, mitotic index, and de-differentiation | Giorgetti et al. (2011) |
| **Other nanoparticles** | | | | | |
| Al$_2$O$_3$ NPs | 13 nm | 2,000 mg/L | Cucumis sativus | Root growth inhibition | Yang and Watts (2005) |
| | – | 1,000–10,000 mg/L | Nicotiana tabacum | Decreased root length, leaf count and biomass | Burklew et al. (2012) |
| NiO NPs | 30 nm | 28–175 mg/L | Lettuce, radish, cucumber | Reduced seed germination and root elongation | Wu et al. (2012) |
| nFe$_{(3)}$O$_{(4)}$ NPs | <50 nm | 400–4,000 mg/L | Arabidopsis thaliana | Reduced seed germination, leaf number, and root elongation | Lee et al. (2010) |
| nSiO$_{(2)}$ NPs | <45 nm | 400–4,000 mg/L | Arabidopsis thaliana | Reduced seed germination, leaf number, and root elongation | Lee et al. (2010) |
| CdSe/ZnS QD | 2–12 nm | 5 mg/L | Arabidopsis thaliana | Oxidative stress | Navarro et al. (2012) |

*MI* mitotic index, *AI* aberration index, *SWCNT* single walled carbon nanotubes, *MWCNT* multiwalled carbon nanotubes; *GR* glutathione reductase (GR, EC 1.6.4.2); *APX* ascorbate peroxidase (APX, EC 1.11.1.11)

Precursor for [Mo6Br14]$^{-2}$     H$_2$O    CMB = Cs$_2$Mo$_6$Br$_{14}$ clusters in Mili–Q water
                                    EtOH   CMB = Cs$_2$Mo$_6$Br$_{14}$ clusters in 95% ethanol

ones. Non-functionalized nanotubes inhibited root elongation in tomato (*Solanum lycopersicum*) plants, whereas enhanced root elongation occurred in onion (*Allium cepa*) and cucumber (*Cucumis sativus*). Root elongation in lettuce (*Lactuca sativa*) was inhibited by functionalized nanotubes, but exposed cabbages (*Brassica oleracea*) and carrots (*Daucus carota*) were unaffected by either form of nanotubes. These effects after CNT exposure tended to be more prominent at a 24-h than at a 48-h incubation. Microscopy images revealed the presence of the nanotube layers on root surfaces, but there was no evidence of uptake (Canas et al. 2008). Nevertheless, this work demonstrated the effect of NP surface properties on phytotoxicity. Results of this study also suggested that plant response to NP exposures depends on the plant species involved, the plant growth stage and nature of the nanomaterial tested.

The toxicity of CNTs also depends on type of dispersion medium in which they are tested. Oleszczuk et al. (2011) evaluated the toxicity of sewage sludge that contained MWCNTs on various plant species. Seed germination and root growth was inhibited. One possible explanation for the inhibition was strong binding of pollutants by the CNTs, thereby inducing plant toxicity. Such a mechanism has been demonstrated in other studies (Yang et al. 2006; Lin and Xing 2008b; Pan and Xing 2008; Oleszczuk et al. 2009).

## 5.2 Titanium Nanoparticles

Evidence available in the Nanowerk nanomaterial database (Database 2013), indicates that the $TiO_2$ nanoparticle (Anatase form), among all other NP categories, is the major type of NP produced worldwide. $TiO_2$ nanoparticles are utilized in paint pigments, paper, ink and in plastics. It is also incorporated into cosmetics such as sunscreens to provide protection against UV light (Larue et al. 2011). Therefore, publications on $TiO_2$ NPs and their interactions with biological entities like plants are increasingly appearing in the literature. Larue et al. (2012) demonstrated that $TiO_2$ NPs accumulate in roots and are distributed throughout wheat plants (*Triticum aestivum* spp.) without being dissolved or having their crystal phase modified. These authors also suggested that there was an upper limit of NP diameter above which no accumulation would occur.

Du et al. (2011) studied the $TiO_2$ NPs under field conditions using a lysimeter. They added the $TiO_2$ NP to soil (0.09 g/kg) and after aging for 2 months, wheat (*Triticum aestivum* L.) was sown. Using TEM (Transmission Electron Microscope) imaging and SEM-X act analysis (X-ray-based detection), $TiO_2$ NPs were observed in primary root tips of wheat plants grown in the presence of $TiO_2$ NPs. Decreased shoot biomass occurred in wheat and provided evidence that the $TiO_2$ NPs caused toxicity. The toxic effect may have been caused by the existence of NPs in cells or their accumulation in cell walls. Other studies indicated that changes wrought by contact with NPs in the microenvironment of the contact area (site of interaction between NP and cell), either from increased metal solubility or from extracellular generation of ROS, could damage cell membranes (Heinlaan et al. 2008; Du et al. 2011). A concentration-dependent abnormality in narbon bean (*Vicia narbonensis* L.) and maize (*Zea mays* L.) was observed by Ruffini et al. (2011). In particular these authors reported effects on the Mitotic Index (MI) and the Aberration Index (AI), which represent expressions of the rate of mitotic cell division and level of chromosomal abnormalities encountered, respectively. Ruffini et al. (2011) reported delayed germination and root elongation from increasing $TiO_2$ NP exposure concentrations. In *Vicia narbonensis* L., the highest tested concentrations (2 and 4%) were required to significantly decrease the MI. In contrast, the MI in *Zea mays* L. was affected at a much lower concentration (0.2%). The increased AI from $TiO_2$ NP exposures were concentration dependent for both plant species. There was also evidence that $TiO_2$ NP caused genotoxicity, characterized primarily by abnormalities rising in the spindle apparatus (c-metaphases, anomalous anaphases, chromosome bridging). Chromosomal damage such as lagging chromosome, fragmentation, micronuclei release, and strand breaks in DNA were also identified, and were similar to changes found in in vivo mouse studies (Trouiller et al. 2009).

Ghosh et al. (2010) observed $TiO_2$ (100 nm) toxicity in *Allium cepa* in the form of DNA damage, growth inhibition and increased lipid peroxidation at a concentration of 319 mg/L. In the same study, $TiO_2$ caused DNA damage to *Nicotiana tabacum*, at a concentration 157 mg/L. Recently, sunscreen-based $TiO_2$ nanocomposites have been studied to determine effects on faba bean (*Vicia faba*) (Anne-Sophie et al. 2011). Results revealed that particles were deposited in the outer root tissue of this plant, possibly from an electrostatic attraction. Such behavior could clog cell pores and interstices, and curtail water circulation and nutrient exchange in the plants. Moreover, deposition in outer root tissue could enhance intake of smaller $TiO_2$ particles by endocytosis via root hairs (Ovecka et al. 2005), or by diffusion into root tissues through the intercellular space without entering cells.

## 5.3  Zinc Nanoparticles

ZnO NPs are one of the most common industrial additives. Like other metal-based NPs, the uptake, translocation, and accumulation of ZnO NPs are poorly understood in plant systems. ZnO NPs are usually dissolved when applied to soil, which enhances plant uptake and phytotoxicity (Du et al. 2011). The way in which ZnO NPs produce toxicity in plants is unclear. Franklin et al. (2007) suggested that ZnO NPs toxicity results from its solubility, whereas Lin and Xing (2008a), studying ryegrass (*Lolium perenne*), indicated that dissolution alone could not be considered as a potential cause of toxicity.

Recently, Lee et al. (2013) studied the effect of ZnO NPs on buckwheat (*Fagopyrum esculentum*) at high concentrations (10–2,000 mg/L). Such exposures caused a biomass drop in buckwheat seedlings, damaged root surface cells, and induced an uncontrolled ROS defense system, i.e., stimulation of catalase activity and antioxidants (Lee et al. 2013).

A RAPD (Random Amplified Polymorphic DNA) analysis was performed by Lopez-Moreno et al. (2010b) to check the effect of ZnO NPs in soybean (*Glycine max*) plants. They found a new DNA band in the RAPD profile of soybean roots that had been treated with ZnO NPs (8 nm) at a 4,000 mg/L concentration. The authors believed that the profile resulted from toxicity due either to the interaction of DNA with zinc ions leached from the ZnO NPs, or from direct interaction with the ZnO NPs. In the same study, XANES (X-ray Absorption Near Edge Structure) spectra from roots treated with 4,000 mg/L ZnO NPs showed the presence of zinc in the oxidized state as Zn (II) within tissues and not as ZnO NPs. No conclusion was reached on how this genotoxic effect had occurred.

## 5.4 Silver Nanoparticles

Silver nanoparticles (Ag NPs) are NPs that are frequently employed in a variety of medical and healthcare products. Their good fit to healthcare derives from their broad-spectrum biocidal properties. Because of their characteristics, these NPs are classified as environmental hazards by the EPA (Environment Protection Agency).

Stampoulis et al. (2009) investigated the effects of Ag NPs and their corresponding bulk counterparts on seed germination, root elongation, and biomass of zucchini (*Cucurbita pepo*). They reported that exposure to Ag NPs at 500 and 100 mg/L resulted in a 57% and 41% decrease in plant biomass and transpiration, respectively, vs. similar measures in controls and plants exposed to bulk Ag. Even shoot accumulation of Ag NPs was an average of 4.7 times greater than for the corresponding bulk solutions. The reason for this may have been that silver NPs produce a greater ion release than does the bulk silver counterpart.

Majumdar and Ahmed (2011) reported that the toxicity of Ag NPs to rice (*Oryza sativa*) was both concentration- and exposure-time-dependent. This conclusion was supported by TEM analysis, which depicted the deposition of silver NPs inside root cells. Those Ag NPs that produced cell wall and vacuole damage primarily had an average diameter of 25 nm. Moreover, Haverkamp and Marshall (2009) observed that silver NPs deposited and accumulated in plant cells did so as a function of the reduction potential present in the system.

After treating water hyssop (*Bacopa monnieri* Linn.) with Ag NPs, mild stress conditions appeared in root, stem and leaf tissue (Krishnaraj et al. 2012). This treatment caused an inconsequential reduction of root and shoot length, disappearance of the air chamber in root cortex, and altered the shape, size, and distribution of xylem elements in stems.

Kumari et al. (2009) treated onion (*Allium cepa*) root tip cells with Ag NPs (diameter <100 nm, 25–100 mg/L) and noted a dose-dependent response. They found a reduced MI frequency from 60.3% (control) to 27.62% (100 mg/L) among silver NP-treated cells. Other chromosomal aberrations also occurred at various silver NP concentrations. Similarly, an in vivo cytogenetic assay carried out by Babu et al. (2008) revealed a dose- and a duration-dependent reduction in MI, along with other chromosomal and mitotic abnormalities in *Allium cepa*. The induction of chromosomal abnormalities indicates that silver NPs has clastogenic potential; clastogenesis is an irreversible genotoxic endpoint in plants. Patlolla et al. (2012) studied the genotoxic effects of silver NPs in *Vicia faba* root-tip, and found that NPs may penetrate the plant system. Once absorbed, silver NPs interfere with intracellular components and cause a dose-dependent decrease in MI and increased chromosomal aberration frequencies.

## 5.5 Copper Nanoparticles

The effect of copper NPs on bioaccumulation and plant seedling growth was investigated by Lee et al. (2008) by using two crop species, mung bean (*Phaseolus radiatus*) and wheat (*Triticum aestivum*). Copper NPs inhibited mung bean (335 mg/L) seedling growth more than wheat (570 mg/L) seedling growth. The higher susceptibility of mung bean seedlings was ascribed to its particular root anatomy and growth architecture.

The first report of root–shoot–root redistribution of CuO NPs (20–40 nm) within maize (*Zea mays* L.) plants was given by Wang et al. (2012). CuO NPs, up to 100 mg/L, had no effect on germination, whereas inhibition of maize seedlings growth occurred. The presence of CuO NPs in xylem sap was detected by TEM and energy dispersive spectroscopy (EDS), proving that CuO NPs are transported from roots to shoots via the xylem. Split-root experiments and high-resolution TEM observations further revealed that CuO NPs were translocated from shoots back to the roots via the phloem, during which Cu (II) was reduced to Cu (I). This study directly confirmed that CuO NPs bioaccumulate and are biotransformed in maize, which increases concern that significant risks may be associated with NPs in food.

## 5.6 Cerium Nanoparticles

Cerium oxide ($CeO_2$) NPs are employed as polishing materials, in fuel cell materials, as additives in glass and ceramics, in agricultural products, and for other uses in the automobile industry (Kosynkin et al. 2000; Corma et al. 2004; Livingston and Helvajian 2005). This rather wide array of uses results in significant releases of $CeO_2$ NPs to the environment. Limbach et al. (2008) estimated the soil concentration of $CeO_2$ NPs to be between 0.32 and 1.12 mg/kg; further, they believed that future releases will continue to rise as time passes.

There are few reports on the toxicity of $CeO_2$ in plants. Lopez-Moreno et al. (2010b) investigated the genotoxicity of $CeO_2$ in soyabean seedlings by using the RAPD assay. The RAPD profile of soyabean roots treated with $CeO_2$ NPs (7 nm) at 2,000 and 4,000 mg/L showed four and three new bands, respectively. The treated plants exhibited an incidence of genetic instability as a result of exposure to these nanoparticles. This genotoxic response was further supported by the appearance of $CeO_2$ NPs in tissues as shown by XANES spectra. In another study with $CeO_2$ NPs,

López-Moreno et al. (2010a) identified reduced germination in corn (30%), tomato (30%), and cucumber (20%) after the plants were treated at 2,000 mg/L.

Zhao et al. (2012) reported how surface coatings and organic matter affected the bioavailability of $CeO_2$ NPs, and studied its uptake mechanism in maize plants (*Zea mays*). Plants were grown in both an unenriched (sandy loam soil) and organic soil (unenriched soil plus high organic matter potting soil in a 1:1 ratio) that had been treated with alginate-coated and uncoated $CeO_2$ NPs. Plants were also exposed to fluorescein isothiocyanate (FITC)-stained $CeO_2$ NPs before being examined under confocal microscopy. In the organic soil, roots treated with uncoated and coated NPs contained more Ce than did roots grown in unenriched soil. In contrast, there was significantly more Ce observed in plant shoots from unenriched soil than in shoots from an organic soil. Confocal fluorescence images revealed the presence of FITC-stained $CeO_2$ NP aggregates in cell walls of the epidermis and cortex of the maize plants that were suggestive of the apoplastic pathway of entry. The presence of $CeO_2$ NP aggregates within vascular tissues was demonstrated by using μXRF (Micro X-Ray Fluorescence) analysis. The tendency of $CeO_2$ NPs to accumulate in plant parts may represent a sizable future threat, i.e., the possibility of entering and moving through the food chain to exhibit toxic responses.

## 5.7 Molybdenum Nanoparticles

Molybdenum NPs are used as catalysts, pigments, corrosion inhibitors, lubricants, etc., and are considered to be naturally toxic unless exposures are low. The phytotoxicity of molybdenum NPs on rapeseed plants (*Brasicca napus*) was tested by Aubert et al. (2012). The form tested by these authors was $Cs_2Mo_6Br_{14}$ (CMB), which provided nanosized hexamolybdenum clusters ($H_2O$-CMB system, $2.3 \pm 0.5$ μm diameter, $390 \pm 60$ nm thickness and 95% Ethanol-CMB system, $550 \pm 180$ nm diameter, $100 \pm 30$ nm thickness) in two systems, water and ethanol. Several toxic effects of these CMB clusters were detected on plants. A concentration-dependent inhibition on rapeseed growth resulted from the treatment. The roots were more affected than the shoots. An exponential increment in root growth inhibition occurred with increasing Mo concentrations. The nanosized Ethanol-CMB cluster treatment produced disturbances in root gravitropism, whereas treatment with $H_2O$-CMB produced high root hair proliferation, and eroded root caps, among other effects. Analysis by surface imaging Nano SIMS (Secondary Ion Mass Spectrometry) showed root morphology damage, perhaps from easier penetration of clusters in the root (Aubert et al. 2012).

## 5.8 Gold Nanoparticles

Gold nanoparticles (Au NPs) are used in or for organic photovoltaics, sensory probes, electronic conductors and catalysis, therapeutic agents, drug delivery, and other biological and medical applications (Peng et al. 2010; Perrault and Chan 2010; Bhatt and Tripathi 2011). Sabo-Attwood et al. (2012) conducted a study to assess the uptake, biodistribution, and toxicity of Au NPs on tobacco plants (*Nicotiana xanthi*). They utilized Synchrotron-based X-ray microanalysis with X-ray absorption near-edge microspectroscopy and high-resolution electron microscopy to localize Au NPs within plants. Results from these experiments revealed that Au NPs entered plants through the roots and moved into the vasculature. Aggregate particles were also detected within the root cell cytoplasm. Furthermore, the Au NP uptake was size selective (viz., 3.5 nm Au NPs were detected in plants but 18 nm Au NPs remained agglomerated on the root outer surfaces). Other effects like leaf necrosis was also observed after 14 days of exposure to 3.5 nm Au NPs. These results generally showed that Au NP entry into plants was size-dependent, and translocation to various tissues occurred to cause phytotoxicity (Sabo-Attwood et al. 2012).

## 5.9 Magnetic Nanoparticles

The impact of magnetic fluid in several plant species has shown significant frequencies of chromosomal aberration and other cytogenetic abnormalities (Pavel et al. 1999; Pavel and Creanga 2005; Racuciu and Creanga 2007). Vochita et al. (2012) reported various genetic effects in sunflower seedling root tip cells from exposure to several magnetic NPs: $ZnFe_2O_4$, $CoFe_2O_4$, and $Fe_3O_4$. Magnetic NP suspensions in the concentration range of 20–100 µl/L produced a 35–50% reduction in MI and an increased AI. Various chromosomal aberrations such as singular or multiple interchromatidian bridges, retard or expulsed chromosomes, chromosome fragments, and micronuclei were also produced. Giorgetti et al. (2011) analyzed the effect of $Fe_3O_4$ NPs on developmental processes of plant by utilizing *Daucus carota* L. in an in vitro model system. $Fe_3O_4$ NPs of 6 nm diameter and a exposure range of 2.01–33.5 mg/L affected the growth, MI and de-differentiation to some extent.

## 5.10 Other Nanoparticles

Several other nanoparticles have been reported to affect plants significantly at different exposure levels. Alumina nanoparticles were toxic to plant species such as *Cucumis sativus* (Yang and Watts 2005) and *Nicotiana tabacum* (Burklew et al. 2012), and affected both root growth and biomass. Wu et al. (2012) assessed the effect of nickel oxide (NiO) and CuO nanoparticles on various plants, measuring seed germination and root elongation effects. Lee et al. (2010) studied *Arabidopsis thaliana* and found that $Fe_3O_4$ and $SiO_2$ nanoparticles significantly affected seed germination and root elongation at various concentrations. The effect of quantum dots (QD) such as CdSe/ZnS (2–12 nm) also caused oxidative stress in *Arabidopsis thaliana* at a very low concentration (Navarro et al. 2012).

## 6 Discussion

The value of fabricated NPs to society has grown rapidly, and so has the emissions from their increased production. Hence, it has become very important to study the impact of NPs on the environment. Although limited ecotoxicity testing protocols have been designed and used, there is still a dearth of methods adequate to study the effect of NPs on plants.

In the last few decades, the toxicological studies performed on nanoparticles in several plant species revealed that not all treated plants manifested toxic effects. However, many researchers believe that the toxicity observed in plants from applying NPs results from plant–nanoparticle physical interactions. The NPs may alter the surface chemistry of the root, thereby affecting the interaction of roots with its environment (Canas et al. 2008). Plant development is negatively affected by exposure to NPs (Asli and Neumann 2009).

In most NP toxicity studies with plants, seed germination and root elongation were used as standard indicators of phytotoxicity, as recommended by the US Environmental Protection Agency. However, seed germination was unaffected by NPs in several studies (Stampoulis et al. 2009), which necessitates finding other toxic endpoints for plants. Plant biomass and chlorophyll levels may well be acceptable phytotoxic endpoints, but need further investigation. Because direct detection and assessment of NP behavior in plants and the environment is difficult, it will be necessary in the future to design more elaborate modeling techniques that better represent and measure the behavior of NPs in biological systems.

Among badly needed research is more detailed information on NP uptake by biological species, NP accumulation in the plant system, and the environmental factors that affect uptake. In addition, more insight is needed on the surface characteristics of roots, shoots, and leaves that significantly affects NP penetration into plants. A predictive model for estimating the toxicity of fabricated NPs is needed, but can only be developed when all major factors that affect mobility, bioaccumulation and cytotoxicity of NPs are sufficiently understood. Any plan that is devised to assess the risks of NPs should, we believe, incorporate four levels of analysis, as depicted in Fig. 3:

1. The assessment should begin with the newly formulated nanoproduct. The product should be screened and analyzed to address the various types of coatings, stabilizers, and matrices that could play a significant role in its environmental safety or behavior.

# Fabricated Nanoparticles: Current Status and Potential Phytotoxic Threats

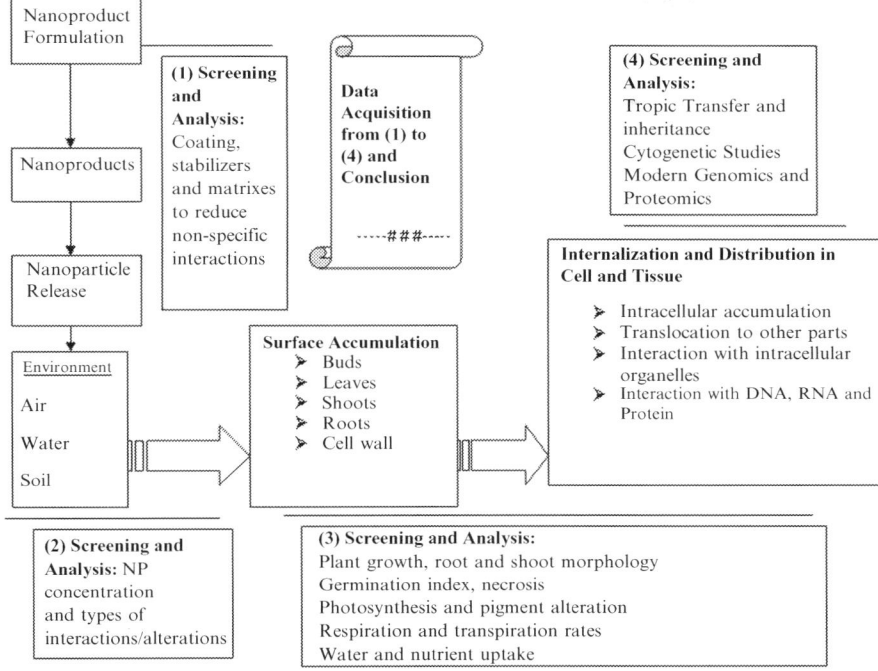

**Fig. 3** A scheme for assessing the risk of nanomaterials to plant systems

2. The nanoparticles released into the environment should be analytically monitored routinely to provide input on environmental concentrations and the nature of chemical modified forms.
3. Plant–nanoparticle interactions should be carefully screened and analyzed to appraise morphological, biochemical, and physiological effects.
4. Modern genomic and proteomic approaches should be utilized to monitor for changes and toxic effects at the cellular and molecular level that may be caused by environmental release of NPs.

The data acquired from undertaking studies at these four levels can be used to broadly describe the environmental behavior of NPs, the core factors affecting phytotoxicity and the mechanism by which it occurs. Having such information will help to draft what will become a suitable risk assessment framework for nanochemicals.

Researchers now have only a limited understanding of how nanoparticles behave in the open environment. Most experts recommend that the manufacturers and retailers that deal with NPs should be responsible for constructing a means to recover and recycle nanomaterials during their product lifecycle. Ideally, nano-based products should be designed for easy separation of NPs from wastes after application, so that they can be isolated and reused.

## 7 Summary

Nanotechnology offers unique attributes to various industrial and consumer sectors, and has become a topic of high interest to scientific communities across the world. Our society has greatly benefitted from nanotechnology already, in that many products with novel properties and wide applicability have been developed and commercialized. However, the increased production and use of nanomaterials have raised concerns about the environmental fate and toxicological implications of nanoparticles and nanomaterials. Research has revealed that various nanomaterials may be hazardous to living organisms. Among biota, plants are widely exposed to released nanomaterials and are sensitive to their effects. The accumulation of nanomaterials in the environment is a potential threat, not only because of potential damage to plants but also because nanoparticles may enter the food chain. Although the literature that addresses the safety of nanoproducts is growing, little is known about the mechanisms by which these materials produce toxicity on natural species, including humans.

In this paper, we have reviewed the literature relevant to what phytotoxic impact fabricated nanoparticles (e.g., carbon nanotubes, metallic and metal oxide nanoparticles, and certain other nanomaterials) have on plants. Nanoparticles produce several effects on plant physiology and morphology. Nanoparticles are known to affect root structure, seed germination, and cellular metabolism. Nanoparticles inhibit growth, induce oxidative stress, morphogenetic abnormalities and produce clastogenic disturbances in several plant species. The size, shape and surface coating of NPs play an important role in determining their level of toxicity. Of course, the dose, route of administration, type of dispersion media, and environmental exposure also contribute to how toxic nanoparticles are to plants.

Currently, nanotoxicity studies are only in their initial phases of development and more research will be required to identify the actual threat nanoproducts pose to the plant system. To date, data show that there is a large variation in the phytotoxicity caused by different NPs. Moreover, the studies conducted thus far have mostly relied on microscopy to detect effects. Studies that incorporate measures and analyses undertaken with more modern tools are needed. Among new data that are most urgently needed on NPs is how fabricated NPs behave once released into the environment, and how exposure to them may affect plant resistance, metabolic pathways, and plant genetic responses.

In this review, we have attempted to collect, present and summarize recent findings from the literature on nanoparticle toxicity in plants. To strengthen the analysis, we propose a scheme for accessing NP toxicity. We also recommend how the potential challenges presented by increased production and release of NPs should be addressed. It is our belief and recommendation that every nanomaterial-based product be subjected to appropriate toxicity and associated assessment before being commercialized.

# References

Anne-Sophie F, Masfaraud JF, Bigorgne E, Nahmani J, Chaurand P, Botta C, Labille J, Rose J, Férard JF, Cotelle S (2011) Environmental impact of sunscreen nanomaterials: ecotoxicity and genotoxicity of altered $TiO_2$ nanocomposites on *Vicia faba*. Environ Pollut 159:2515–2522

ANUI (2012) Apply nanotech to up industrial, agri output. The Daily Star (Bangladesh). http://www.thedailystar.net/newDesign/news-details.php?nid=230436

Asli S, Neumann M (2009) Colloidal suspensions of clay or titanium dioxide nanoparticles can inhibit leaf growth and transpiration via physical effects on root water transport. Plant Cell Environ 32:577–584

Aubert T, Burel A, Esnault MA, Cordier S, Grasset F, Cabello-Hurtado F (2012) Root uptake and phytotoxicity of nanosized molybdenum octahedral clusters. J Hazard Mater 219–220:111–118

Auffan M, Bottero JY, Chaneac C, Rose J (2010) Inorganic manufactured nanoparticles: how their physicochemical properties influence their biological effects in aqueous environments. Nanomedicine 5(6):999–1007

Babu K, Deepa M, Shankar SG, Rai S (2008) Effect of nano-silver on cell division and mitotic chromosomes: a prefatory siren. Internet J Nanotechnol 2(2):2. doi:10.5580/10eb

Barrena R, Casals E, Colan J, Font X, Sanchez A, Puntes V (2009) Evaluation of the ecotoxicology of model nanoparticles. Chemosphere 75:850–857

Bhatt I, Tripathi BN (2011) Interaction of engineered nanoparticles with various components of the environment and possible strategies for their risk assessment. Chemosphere 82:308–317

Bosetti M, Mass A, Tobin E, Cannas M (2002) Silver coated materials for external fixation devices: *in vitro* biocompatibility and genotoxicity. Biomaterials 23:887–892

Burklew CE, Ashlock J, Winfrey WB, Zhang B (2012) Effects of aluminum oxide nanoparticles on the growth, development, and microRNA expression of tobacco (*Nicotiana tabacum*). PLoS One 7(5):e34783

Buzea C, Pacheco II, Robbie K (2007) Nanomaterials and nanoparticles: sources and toxicity. Biointerphases 2(4):17–71

Cabiscol E, Tamarit J, Ros J (2000) Oxidative stress in bacteria and protein damage by reactive oxygen species. Int Microbiol 3:3–8

Canas JE, Long M, Nations S, Vadan R, Dai L, Luo M, Ambikapathi R, Lee EH, Olszyk D (2008) Effects of functionalized and non-functionalized single-walled carbon nanotubes on root elongation of select crop species. Environ Toxicol Chem 27:1922–1931

Capala J, Barth RF, Bendayam M, Lauzon M, Adams DM, Soloway AH, Fenstermaker RA, Carlsson J (1996) Boronated epidermal growth factor as a potential targeting agent for boron neutron capture therapy of brain tumors. Bioconjug Chem 7:7–15

Cheng XK, Kan AT, Tomsom MB (2004) Naphthalene adsorption and desorption from aqueous C-60 fullerene. J Chem Eng Data 49:675–683

Cho M, Chung H, Choi W, Yoon J (2005) Different inactivation behaviors of MS-2 phage and *Escherichia coli* in $TiO_2$ photocatalytic disinfection. Appl Environ Microbiol 71(1):270–275

Corma A, Atienzar P, Garcia H, Chane-Ching JY (2004) Hierarchically mesostructured doped $CeO_2$ with potential form solar-cell use. Nat Mater 3:394–397

Database (2013) Nanowerk Nanomaterial Database [Internet]. Available from: http://www.nanowerk.com/phpscripts/n_dbsearch.php

Du W, Sun Y, Ji R, Zhu J, Wu J, Guo H (2011) $TiO_2$ and ZnO nanoparticles negatively affect wheat growth and soil enzyme activities in agricultural soil. J Environ Monit 13:822–828

Eichert T, Kurtz A, Steiner U, Goldbach HE (2008) Size exclusion limits and lateral heterogeneity of the stomatal foliar uptake pathway for aqueous solutes and water-suspended nanoparticles. Physiol Plant 134:151–160

El Nemr A, Abd-Allah AMA (2003) Contamination of polycyclic aromatic hydrocarbons (PAHs) in microlayer and subsurface waters along Alexandria coast, Egypt. Chemosphere 52:1711–1716

Farre M, Sanchis J, Barcelo D (2011) Analysis and assessment of the occurrence, the fate and the behavior of nanomaterials in the environment. Trends Anal Chem 30(3):517–527

Fernandez V, Eichert T (2009) Uptake of hydrophilic solutes through plant leaves: current state of knowledge and perspectives of foliar fertilization. Crit Rev Plant Sci 28:36–68

Foraker AB, Walczak RJ, Cohen MH (2003) Microfabricated porous silicon particles enhance paracellular delivery of insulin across intestinal Caco-2 cell monolayers. Pharm Res 20:110–116

Franklin NM, Rogers NJ, Apte SC, Batley GE, Gadd GE, Casey PS (2007) Comparative toxicity of nanoparticulate ZnO, bulk ZnO, and $ZnCl_2$ to a freshwater microalga (*Pseudokirchneriella subcapitata*): the importance of particle solubility. Environ Sci Technol 41(24):8484–8490

García A, Espinosa R, Delgado L, Casals E, González E, Puntes V, Barata C, Font X, Sánchez A (2011) Acute toxicity of cerium oxide, titanium oxide and iron oxide nanoparticles using standardized tests. Desalination 269:136–141

Ghosh M, Bandyopadhyay M, Mukherjee A (2010) Genotoxicity of titanium dioxide $TiO_2$ nanoparticles at two trophic levels: plant and human lymphocytes. Chemosphere 81:1253–1262

Giorgetti L, Ruffini Castiglione M, Bernerbini M, Geri C (2011) Nanoparticles effects on growth and differentiation in cell culture of carrot—*Daucus carota L.*. Agrochimica LV:45–53

Gotovac S, Honda H, Hattori Y, Takahashi K, Kanoh H, Kaneko K (2007) Effect of nanoscale curvature of single-walled carbon nanotubes on adsorption of polycyclic aromatic hydrocarbons. Nano Lett 7:583–587

Gupta VK, Rastogi A (2008) Biosorption of lead from aqueous solution by green algae *Spirogyra* species: kinetic and equilibrium studies. J Hazard Mater 152(1):407–414

Gupta SM, Tripathi M (2011) A review of $TiO_2$ nanoparticles. Chin Sci Bull 56:1639–1657

Haverkamp RG, Marshall AT (2009) The mechanism of metal nanoparticle formation in plants: limits on accumulation. J Nanopart Res 11:1453–1463

Heinlaan M, Ivask A, Blinova I, Dubourguier H, Kahru A (2008) Toxicity of nanosized and bulk ZnO, CuO and $TiO_2$ to bacteria *Vibrio fischeri* and crustaceans *Daphnia magna* and *Thamnocephalus platyurus*. Chemosphere 71:1308–1316

Hoshino K, Gopal A, Glaz M, Bout DV, Zhang XJ (2012) Nanoscale fluorescence imaging with quantum dot near-field electroluminescence. Appl Phys Lett 101(2–3):043118

Howarth M, Liu W, Puthenveetil S, Zheng Y, Marshall LF, Schmidt MM, Wittrup KD, Bawendi MG, Ting AY (2008) Monovalent, reduced-size quantum dots for imaging receptors on living cells. Nat Methods 5(5):397–399

Hu X, Liu J, Mayer P, Jiang G (2008) Impacts of some environmentally relevant parameters on the sorption of polycyclic hydrocarbons to aqueous suspensions of fullerene. Environ Toxicol Chem 27(9):1868–1874

Huber DL (2005) Synthesis, properties, and applications of iron nanoparticles. Small 1(5):482–501

Jia G, Wang HF, Yan L, Wang X, Pei RJ, Yan T, Zhao YL, Guo XB (2005) Cytotoxicity of carbon nanomaterials: single-wall nanotube, multiwall nanotube, and fullerene. Environ Sci Technol 39:1378–1383

Khodakovskaya M, Dervishi E, Mahmood M, Xu Y, Li Z, Watanabe F, Biris AS (2009) Carbon nanotubes are able to penetrate plant seed coat and dramatically affect seed germination and plant growth. ACS Nano 3:3221–3227

Khus M, Gernjak W, Ibanez PF, Rodriguez SM, Galvez JB, Icli S (2006) A comparative study of supported $TiO_2$ as photocatalyst in water decontamination at solar pilot plant scale. J Sol Energy 128:331–337

Klaine SJ, Alvarez PJJ, Batley GE, Fernandes TF, Handy RD, Lyon DY, Mahendra S, McLaughlin MJ, Lead JR (2008) Nanomaterials in the environment: behavior, fate, bioavailability, and effects. Environ Toxicol Chem 27(9):1825–1851

Klajnert B, Bryszewska M (2001) Dendrimers: properties and applications. Acta Biochim Pol 48(1):199–208

Konstantatos G, Sargent EH (2009) Solution-processed quantum dot photodetectors. Proc IEEE 97(10):1666–1683

Kosynkin VD, Arzgatkina AA, Ivanov EN, Chtoutsa MG, Grabko AI, Kardapolov AV, Sysina NA (2000) The study of process production of polishing powder based on cerium dioxide. J Alloys Compd 303–304:421–425

Krishnaraj C, Jagan EG, Ramachandran R, Abirami SM, Mohan N, Kalaichelvan PT (2012) Effect of biologically synthesized silver nanoparticles on *Bacopa monnieri (Linn.)* Wettst. plant growth metabolism. Process Biochem 47:651–658

Kukowska-Latallo JF, Raczka E, Quintana A, Chen CL, Rymaszewski M, Baker JR (2000) Intravascular and endobronchial DNA delivery to murine lung tissue using a novel, nonviral vector. Hum Gene Ther 11:1385–1395

Kumar V, Kumari A, Guleria P, Yadav SK (2012) Evaluating the toxicity of selected types of nanochemicals. Rev Environ Contamin Toxicol 215:39–121

Kumari M, Mukherjee A, Chandrasekaran N (2009) Geno-toxicity of silver nanoparticles in *Allium cepa*. Sci Total Environ 407:5243–5245

Larue C, Khodja H, Herlin-Boime N, Brisset F, Flank AM, Fayard B, Chaillou S, Carriere M (2011) Investigation of titanium dioxide nanoparticles toxicity and uptake by plants. J Phys Conf Ser 304(1):012057

Larue C, Laurette J, Herlin-Boime N, Khodja H, Fayard B, Flank AM, Brisset F, Carriere M (2012) Accumulation, translocation and impact of $TiO_2$ nanoparticles in wheat (*Triticum aestivum spp.*): influence of diameter and crystal phase. Sci Total Environ 431:197–208

Lee SH, Richards RJ (2004) Montserrat volcanic ash induces lymph node granuloma and delayed lung inflammation. Toxicology 195:155–165

Lee W, An Y, Yoon H, Kweon H (2008) Toxicity and bioavailability of copper nanoparticles to the terrestrial plants mung bean (*Phaseolus radiatus*) and wheat (*Triticum aestivum*): plant uptake for water insoluble nanoparticles. Environ Toxicol Chem 27(9):1915–1921

Lee CW, Mahendra S, Zodrow K, Li D, Tsai YC, Braam J, Alvarez PJ (2010) Developmental phytotoxicity of metal oxide nanoparticles to *Arabidopsis thaliana*. Environ Toxicol Chem 3:669–675

Lee S, Kim S, Kim S, Lee I (2013) Assessment of phytotoxicity of ZnO NPs on a medicinal plant, *Fagopyrum esculentum*. Environ Sci Pollut Res Int 20(2):848–854

Limbach LK, Bereiter R, Muller E, Krebs R, Galli R, Stark WJ (2008) Removal of oxide nanoparticles in a model wastewater treatment plant: influence of agglomeration and surfactants on clearing efficiency. Environ Sci Technol 42:5828–5833

Lin D, Xing B (2007) Phytotoxicity of nanoparticles: inhibition of seed germination and root growth. Environ Pollut 150:243–250

Lin D, Xing B (2008a) Root uptake and phytotoxicity of ZnO nanoparticles. Environ Sci Technol 42:5580–5585

Lin DH, Xing B (2008b) Adsorption of phenolic compounds by carbon nanotubes: role of aromaticity and substitution of hydroxyl groups. Environ Sci Technol 42:7254–7259

Lin C, Fugetsu B, Su Y, Watari F (2009a) Studies on toxicity of multi-walled carbon nanotubes on *Arabidopsis* T87 suspension cells. J Hazard Mater 170:578–583

Lin S, Reppert J, Hu Q, Hudson JS, Reid ML, Ratnikova TA, Rao AM, Luo H, Ke PC (2009b) Uptake, translocation, and transmission of carbon nanomaterials in rice plants. Small 5:1128–1132

Livingston FE, Helvajian H (2005) Variable UV laser exposure processing of photosensitive glass–ceramics: maskless micro to meso-scale structure fabrication. Appl Phys A 81:1569–1581

López-Moreno ML, de la Rosa G, Hernández-Viezcas JA, Peralta-Videa JR, Gardea-Torresdey JL (2010a) XAS Corroboration of the uptake and storage of $CeO_2$ nanoparticles and assessment of their differential toxicity in four edible plant species. J Agric Food Chem 58(6):3689–3693

Lopez-Moreno ML, De La Rosa G, Hernandez-Viezcas JA, Castillo-Michel H, Botez CE, Peralta-Videa JR, Gardea-Torresdey JL (2010b) Evidence of the differential biotransformation and genotoxicity of ZnO and $CeO_2$ nanoparticles on soybean (*Glycine max*) plants. Environ Sci Technol 44:7315–7320

Ma X, Lee JG, Deng Y, Kolmakov A (2010) Interactions between engineered nanoparticles (ENPs) and plants: phytotoxicity, uptake and accumulation. Sci Total Environ 408(16):3053–3061

Mahajan P, Dhoke SK, Khanna AS (2011) Effect of nano-ZnO particle suspension on growth of Mung (*Vigna radiata*) and Gram (*Cicer arietinum*) seedlings using plant agar method. J Nanotechnol 2011:1–7

Majumdar H, Ahmed GU (2011) Phytotoxicity effect of silver nanoparticles on *Oryza sativa*. Int J ChemTech Res 3(3):1494–1500

Moaveni P, Karimi K, Zare Valojerdi M (2011) The nanoparticles in plants: review paper. J Nanostruct Chem 2(1):59–78

Moore MN (2006) Do nanoparticles present ecotoxicological risks for the health of the aquatic environment? Environ Int 32:967–976

Murr LE, Esquivel EV, Bang JJ, de la Rosa G, Gardea-Torresdey JL (2004) Chemistry and nanoparticulate compositions of a 10,000 year-old ice core melt water. Water Resour 38:4282–4296

Nair R, Varghese SH, Nair BG, Maekawa T, Yoshida Y, Kumar DS (2010) Nanoparticulate material delivery to plants. Plant Sci 179:154–163

Navarro E, Baun A, Behra R, Hartmann NB, Filser J, Miao AJ, Quigg A, Santschi PH, Sigg L (2008a) Environmental behaviour and ecotoxicology of engineered nanoparticles to algae, plant and fungi. Environ Sci Technol 17:372–386

Navarro E, Piccipetra F, Wagner B, Marconi F, Kaegi R, Odzak N, Sigg L, Behra R (2008b) Toxicity of silver nanoparticles to *Chlamydomonas reinhardtii*. Environ Sci Technol 42:8959–8964

Navarro DA, Bisson MA, Agaa DS (2012) Investigating uptake of water-dispersible CdSe/ZnS quantum dot nanoparticles by *Arabidopsis thaliana* plants. J Hazard Mater 211–212:427–435

Novack B, Bucheli TD (2007) Occurrence, behavior and effects of nanoparticles in the environment. Environ Pollut 150:5–22

O'Farrell N, Houlton A, Horrocks BR (2006) Silicon nanoparticles: applications in cell biology and medicine. Int J Nanomed 1(4):451–472

Oleszczuk P, Pan B, Xing B (2009) Adsorption and desorption of oxytetracycline and carbamazepine by multiwalled carbon nanotubes. Environ Sci Technol 43:9167–9173

Oleszczuk P, Jósko I, Xing B (2011) The toxicity to plants of the sewage sludges containing multiwalled carbon nanotubes. J Hazard Mater 186:436–442

Ovecka M, Lang I, Baluska F, Ismail A, Illes P, Lichtscheidl IK (2005) Endocytosis and vesicle trafficking during tip growth of root hairs. Protoplasma 226(1–2):39–54

Pan B, Xing B (2008) Adsorption mechanisms of organic chemicals on carbon nanotubes. Environ Sci Technol 42:9005–9013

Patlolla AK, Berry A, May L, Tchounwou PB (2012) Genotoxicity of silver nanoparticles in *Vicia faba*: a pilot study on the environmental monitoring of nanoparticles. Int J Environ Res Publ Health 9:1649–1662

Pavel A, Creanga DE (2005) Chromosomal aberrations in plants under magnetic fluid influence. J Magn Magn Mater 289:469–472

Pavel A, Trifan M, Bara II, Creanga DE, Cotae C (1999) Accumulation dynamics and some cytogenetical tests at *Chelidonium majus* and *Papaver somniferum* callus under the magnetic liquid effect. J Magn Magn Mater 201(1–3):443–445

Peng G, Hakim M, Broza YY, Billan S, Abdah-Bortnyak R, Kuten A, Tisch U, Haick H (2010) Detection of lung, breast, colorectal, and prostate cancers from exhaled breath using a single array of nanosensors. Br J Cancer 103(4):542–551

Perrault SD, Chan WCW (2010) *In vivo* assembly of nanoparticle components to improve targeted cancer imaging. Proc Natl Acad Sci U S A 107:11194–11199

Pulickel MA, Zhou OZ (2001) Applications of carbon nanotubes. Top Appl Phys 80:391–425

Racuciu M, Creanga DE (2007) Cytogenetic changes induced by aqueous ferrofluids in agricultural plants. J Magn Magn Mater 311(1):288–291

Reid BJ, Jones KC, Semple KT (2000) Bioavailability of persistent organic pollutants in soils and sediments—a perspective on mechanisms, consequences and assessment. Environ Pollut 108:103–112

Remedios C, Rosario F, Bastos V (2012) Environmental nanoparticles interactions with plants: morphological, physiological, and genotoxic aspects. J Bot 2012:1–8

Rico CM, Majumdar S, Duarte-Gardea M, Peralta-Videa JR, Gardea-Torresdey JL (2011) Interaction of nanoparticles with edible plants and their possible implications in the food chain. J Agric Food Chem 59(8):3485–3498

Rietmeijer FJM, Mackinnon IDR (1997) Bismuth oxide nanoparticles in the stratosphere. J Geophys Res E 102:6621–6627

Roco MC (2003) Nanotechnology: convergence with modern biology and medicine. Curr Opin Biotechnol 14:337–346

Roy R, Zanini D, Meunier SJ, Romanowska A (1993) Solid-phase synthesis of dendritic sialoside inhibitors of influenza A virus haemagglutinin. J Chem Soc Chem Commun 1869–1872

Royal Society (2004) Nanoscience and nanotechnologies: opportunities and uncertainties. Report by the Royal Society and the Royal Academy of Engineering. http://www.nanotec.org.uk/finalReport.htm

Rubasinghe G, Elzey S, Baltrusaitis J, Jayaweera PM, Grassian VH (2010) Reactions on atmospheric dust particles: surface photochemistry and size-dependent nanoscale redox chemistry. J Phys Chem Lett 1:1729–1737

Ruffini Castiglione M, Cremonini R (2009) Nanoparticles and higher plants. Caryologia 62(2):161–165

Ruffini Castiglione M, Geri C, Giorgetti L, Cremonini R (2011) The effects of nano-$TiO_2$ on seed germination, development and mitosis of root tip cells of *Vicia narbonensis L.* and *Zea mays L.*. J Nanopart Res 13:2443–2449

Sabo-Attwood T, Unrine JM, Stone JW, Murphy CJ, Ghoshroy S, Blom D, Bertsch PM, Newman LA (2012) Uptake, distribution and toxicity of gold nanoparticles in tobacco (*Nicotiana xanthi*) seedlings. Nanotoxicology 6(4):353–360

SCENIHR (Scientific Committee on Emerging and Newly Identified Health Risks) (2005) The appropriateness of existing methodologies to assess the potential risks associated with engineered and adventitious products of nanotechnologies (SCENIHR report 002/05) (European Commission: Scientific Committee on Emerging and Newly Identified Health Risks). http://ec.europa.eu/health/ph_risk/committees/04_scenihr/docs

Schmid K, Riediker M (2008) Use of nanoparticles in Swiss industry: a targeted survey. Environ Sci Technol 42(7):2253–2260

Shahmoradi B, Ibrahim IA, Sakamoto N, Ananda S, Somashekar R, Guru Row TN, Byrappa K (2010) Photocatalytic treatment of municipal wastewater using modified neodymium doped $TiO_2$ hybrid nanoparticles. J Environ Sci Health A 45:1248–1255

Shen CX, Zhang QF, Li J, Bi FC, Yao N (2010) Induction of programmed cell death in *Arabidopsis* and rice by single-wall carbon nanotubes. Am J Bot 97:1–8

Smijs TG, Pavel S (2011) Titanium dioxide and zinc oxide nanoparticles in sunscreens: focus on their safety and effectiveness. Nanotechnol Sci Appl 4:95–112

Smirnova EA, Gusev AA, Zaitseva ON, Lazareva EM, Onishchenko GE, Kuznetsova EV, Tkachev AG, Feofanov AV, Kirpichnikov MP (2011) Multi-walled carbon nanotubes penetrate into plant cells and affect the growth of *Onobrychis arenaria* seedlings. Acta Nat 3(1):99–106

Smita S, Gupta SK, Bartonova A, Dusinska M, Gutleb AC, Rahman Q (2012) Nanoparticles in the environment: assessment using the causal diagram approach. Environ Health 11(Suppl 1):S13, 10.1186/1476-069X-11-S1-S13

Somasundaran P, Fang X, Ponnurangam S, Li B (2010) Nanoparticles: characteristics, mechanisms and modulation of biotoxicity. KONA Powder Part J 28:38–49

Srividya K, Mohanty K (2009) Biosorption of hexavalent chromium from aqueous solutions by *Catla catla* scale: equilibrium and kinetics studies. Chem Eng J 155:666–673

Stampoulis D, Sinha SK, White JC (2009) Assay-dependent phytotoxicity of nanoparticles to plants. Environ Sci Technol 43:9473–9479

Tan XM, Fugetsu B (2007) Multi-walled carbon nanotubes interact with cultured rice cells: evidence of a self-defense response. J Biomed Nanotechnol 3:285–288

Tan XM, Lin C, Fugetsu B (2009) Studies on toxicity of multi-walled carbon nanotubes on suspension rice cells. Carbon 47:3479–3487

Trouiller B, Reliene R, Westbrook A, Solaimani P, Schiestl RH (2009) Titanium dioxide nanoparticles induce DNA damage and genetic instability *in vivo* in mice. Cancer Res 69:8784–8789

Twyman LJ, Beezer AE, Esfand R, Hardy MJ, Mitchell JC (1999) The synthesis of water soluble dendrimers, and their application as possible drug delivery systems. Tetrahedron Lett 40:1743–1746

Unfried K, Albrecht C, Klotz LO, Mikecz A, Grether-Beck S, Schin RPF (2007) Cellular responses to nanoparticles: target structures and mechanisms. Nanotoxicology 1(1):52–71

USEPA (2007) Nanotechnology white paper. Document Number EPA 100/B-07001. http://www.epa.gov/osa

Uzu G, Sobanska S, Sarret G, Munoz M, Dumat C (2010) Foliar lead uptake by lettuce exposed to atmospheric pollution. Environ Sci Technol 44:1036–1042

Verma HC, Upadhyay C, Tripathi A, Tripathi RP, Bhandari N (2002) Thermal decomposition pattern and particle size estimation of iron minerals associated with the cretaceous-tertiary boundary at Gubbio. Meteorit Planet Sci 37:901–909

Vochita G, Creanga D, Focanici-Ciurlica EL (2012) Magnetic nanoparticle genetic impact on root tip cells of sunflower seedlings. Water Air Soil Pollut 223:2541–2549

Wang S, Kurepa J, Smalle JA (2011) Ultra-small $TiO_2$ nanoparticles disrupt microtubular networks in *Arabidopsis thaliana*. Plant Cell Environ 34(5):811–820

Wang Z, Xie X, Zhao J, Liu X, Feng W, White JC, Xing B (2012) Xylem- and phloem-based transport of CuO nanoparticles in maize (*Zea mays L.*). Environ Sci Technol 46(8):4434–4441

Wigginton NS, Haus KL, Hochella MF (2007) Aquatic environmental nanoparticles. J Environ Monit 9:1306–1316

Wu SG, Huang L, Head J, Chen DR, Kong IC, Tang YJ (2012) Phytotoxicity of metal oxide nanoparticles is related to both dissolved metal ions and adsorption of particles on seed surfaces. J Pet Environ Biotechnol 3(4):126

Yang L, Watts J (2005) Particle surface characteristics may play an important role in phytotoxicity of alumina nanoparticles. Toxicol Lett 158:122–132

Yang K, Zhu L, Xing B (2006) Adsorption of polycyclic aromatic hydrocarbons by carbon nanomaterials. Environ Sci Technol 40(6):1855–1861

Yeo SY, Lee HJ, Jeong SH (2003) Preparation of nanocomposite fibers for permanent antibacterial effect. J Mater Sci 38:2143–2147

Yu-Nam Y, Lead R (2008) Manufactured nanoparticles: an overview of their chemistry, interactions and potential environmental implications. Sci Total Environ 400:396–414

Zhao XU, Liz W, Chen Y, Ahi LY, Zhu YF (2007) Solid-phase photocatalytic degradation of polyethylene plastic under UV and solar light irradiation. J Mol Catal A Chem 268:101–106

Zhao L, Peralta-Videa JR, Varela-Ramirez A, Castillo-Michel H, Li C, Zhang J, Aguilera RJ, Keller AA, Gardea-Torresdey JL (2012) Effect of surface coating and organic matter on the uptake of $CeO_2$ NPs by corn plants grown in soil: insight into the uptake mechanism. J Hazard Mater 225–226:131–138

# Status of Heavy Metal Residues in Fish Species of Pakistan

Majid Hussain, Said Muhammad, Riffat N. Malik,
Muhammad U. Khan, and Umar Farooq

## Contents

1 Introduction .................................................................................................... 111
2 Heavy Metal Residues in Fish Species ........................................................... 113
   2.1 Freshwater Fish ........................................................................................ 113
   2.2 Marine Fish .............................................................................................. 124
   2.3 Heavy Metal Residues in Fish Species of Neighboring Countries ......... 126
3 Conclusions and Recommendations ................................................................ 128
4 Summary ......................................................................................................... 128
References ............................................................................................................ 129

## 1 Introduction

Heavy metals (HM) are considered to be dangerous because of their toxicity and natural persistence, and pollution by them in recent decades has become a global issue (Vuren et al. 1999; Shahid et al. 2011; Shah et al. 2012). These HM are concentrated

---

M. Hussain
Department of Forestry and Wildlife Management, University of Haripur,
Hattar Road Haripur, Khyber Pakhtunkhwa, Pakistan

S. Muhammad (✉)
Key Laboratory of Alpine Ecology and Biodiversity, Institute of Tibetan Plateau Research,
Chinese Academy of Sciences, Beijing 100101, PR China

Department of Earth Sciences, COMSATS Institute of Information Technology (CIIT),
Abbottabad, Pakistan
e-mail: saidmuhammad1@gmail.com

R.N. Malik • M.U. Khan
Environmental Biology and Ecotoxicology Laboratory, Department of Environmental Sciences,
Quaid-i-Azam University, Islamabad 45320, Pakistan

U. Farooq
Department of Physics, Islamic International University, Islamabad 44000, Pakistan

in certain environmental compartments such as soil, sediment, tailing deposits, and wastewater (Demirak et al. 2006; Malik et al. 2011; Muhammad et al. 2011a, 2013; Hajeb et al. 2012; Shah et al. 2013; Shahid et al. 2013). As a result of various natural (i.e., weathering and erosion of bed rocks and ore deposits) and anthropogenic (i.e., urban, industrial, mining, and agricultural) activities, HM released into the environment (Zhang et al. 2008; Muhammad et al. 2010, 2011b). HM are mobile in the environment and ultimately find their way to aquatic ecosystems (Javid 2005; Demirak et al. 2006; Swaileh and Sansur 2006; Shah et al. 2012). Fresh water ecosystems (e.g., streams, rivers, and lakes) are severely affected by HM contamination (Paul and Meyer 2001; Arian et al. 2008; Muhammad et al. 2010, 2011b).

Heavy metals represented by some examples that are severe toxicants (As, Hg, Cd, and Pb), and these may alter water quality, cause adverse effects, and can structurally modify aquatic life, especially fish (Chang et al. 1998; Shah et al. 2012; Khan et al. 2012). However, when fish take these toxicants in, they are sometimes metabolized to even more toxic derivatives (Duffus 1980). For example, mercury is microbially converted into methyl mercury, which is much more bioavailable and more toxic than metallic mercury (Dix 1981).

Among aquatic biota, certain fish species are top consumers (Dallinger et al. 1987). Fish contact HM in four major ways: from consuming food, via direct water uptake (gills), from consuming nonedible particles, and via skin absorption. Once absorbed, HM enter the bloodstream and ultimately are carried to various fish organs like liver, kidneys, and gills prior to being eliminated or stored (Nussey et al. 2000). Intake of HM by fish may reduce food utilization and ultimately reduce metabolic rates. As a result, the skin, muscles, liver, kidneys, and other tissues are affected in ways that hamper growth and development (Javid 2005). Once fish absorb HM, their concentrations in muscle remain nearly constant for life (Rashed 2001). However, the degree to which fish store and bioaccumulate HM depends on both absorption and elimination rates in different body organs.

Assessing and monitoring HM levels that exist in fish muscles gives a direct measure of how much metal may be transferred to humans that consume fish, and therefore, of the potential subsequent effects on human health. Moreover, assessing the HM content of fish tissues may also provide input on the environmental status of aquatic ecosystems (Widianarko et al. 2000). Hence, fish may act as a bioindicator from which one can determine the extent to which HM contaminate an aquatic ecosystem (Javid 2003).

Pakistan has a number of rivers (Indus, Jhelum, Chenab, Kabul, etc.) and lakes (Mancher, Keenjhar, Rawal, etc.). Among these the Indus River represents one of the largest water distribution systems in South East Asia (Ittekkot and Arian 1986). Unfortunately, rivers worldwide too frequently receive waste from bedrock and ore deposits, mining activities, tailing deposits, industry effluent, and solid waste and wastewater disposal that impair water quality and threaten aquatic life and human health (Jaffar et al. 1988; Jaleel et al. 1991; Ashraf et al. 1991; Muhammad et al. 2010, 2011b; Khan et al. 2011). During the last few decades the human population in many areas has increased many times, and this has produced both rapid urbanization and industrialization. Such urbanization and industrialization utilizes more natural resources to fulfill human needs and produces huge quantities of solid and liquid wastes that includes hazardous HM that are dumped into the environment. Ultimately hazardous HM reach and contaminate natural waters and fish (Qadir and Malik 2011; Khan et al. 2012). In Pakistan, the contamination by HM of aquatic ecosystems is reaching alarming levels, despite the fact that the Indus River and its tributaries serve to feed and water millions of people (Tariq et al. 1996; Javid 2005; Qadir et al. 2008). Such examples justify the attention given to HM contamination of water and fish by environmental scientists in recent years (Fatoki and Mathabatha 2001; Qadir and Malik 2011; Khan et al. 2012; Cardwell et al. 2013).

To date, no comprehensive review of HM contamination of fish has been performed in Pakistan. The studies that have been conducted on HM residues in fish species of Pakistan were limited either to a single catch site or to a specific HM or fish species. Therefore, it is our aim in this study to summarize and evaluate the results of HM concentrations in fish species of Pakistan, both to provide a clear picture of the current status of this topic and to suggest future research that is needed to fill the gaps in this topic area.

## 2 Heavy Metal Residues in Fish Species

Globally, extensive research has been carried out on HM contamination of and uptake by fish species. The main body of research has addressed the biological effects of the HM focusing on endpoints such as immunotoxicity, carcinogenicity, teratogenicity, and neurotoxicity. In Pakistan, several researchers have reported the distribution and sources of HM contamination in fish and degree of uptake by fish species (Table 1 and Fig. 1). HM residues in fish species vary significantly from location to location (Fig. 2). In this article, we place emphasis on reviewing the HM contaminations of fish species collected at different locations in Pakistan.

### *2.1 Freshwater Fish*

Freshwater reservoirs (i.e., rivers and lakes) in Pakistan provide drinking water to local inhabitants and for irrigating cropland (Qadir et al. 2008). The streams that drain the main rivers and flow across the alluvial plains of Pakistan are locally called Nullahs. These Nullahs have been become contaminated with a large quantity of untreated industrial effluents and household sewage, which ultimately deteriorates water quality and the involved ecosystems (Malik and Nadeem 2010; Khan et al. 2012). These Nullahs are known to be contaminated by Cr (18.20 mg/L), Ni (1.29 mg/L), Mn (4.88 mg/L), Pb (7.32 mg/L), and Cd (0.64 mg/L), and these HM may concentrate in fish and in other environmental compartments (Qadir and Malik 2011; Khan et al. 2013). Below, we address the degree to which HM residues have contaminated fish species of different provinces of Pakistan.

**Table 1** Heavy metal concentrations (µg/g) determined in freshwater and marine fish species studied during 1990–2012

| River/Lake | Site | Organs | Zn | Cu | Cd | Cr |
|---|---|---|---|---|---|---|
| Khyber Pakhtunkhwa Province | | | | | | |
| Kabul River (15) | Nowshera/Warsak | Gills | 1,224.90–2.414.00 | 67.70–76.70 | | 6.02–6.60 |
| Kabul River (54) | Nowshera | Skin | 995.00–1,971.00 | 89.00–207.00 | 63.00–69.70 | 525.30–709.70 |
| | | Gills | 886.00–1,618.70 | 97.30–167.00 | 71.00–74.00 | 600.00–730.30 |
| | | Intestine | 470.00–982.70 | 101.30–293.00 | 62.30–69.00 | 451.00–870.30 |
| | | Liver | 509.00–1,175.70 | 136.00–1,644.00 | 64.30–72.30 | 513.00–643.70 |
| | | Muscle | 649.00–883.00 | 46.30–191.70 | 66.70–68.00 | 533.30–647.30 |
| Kabul River | Nowshera | BT[a] | 826.00–8,391.30 | 114.70–159.00 | 53.30–66.0 | 489.00–570.00 |
| Shah Alam | Peshawar | Gills | 0.23–1.56 | 0.09–2.30 | | |
| (Kabul River) (12) | | Liver | 0.32–1.56 | 0.08–0.88 | | |
| | | Muscle | 0.06–0.32 | 0.21–0.86 | | |
| Punjab Province | | | | | | |
| Indus River (164) | Tarbela | Muscles | 1.49–5.58 | 0.42–0.81 | 0.03–0.09 | 0.31–0.91 |
| | Chashma | Muscles | 1.31–2.17 | 0.21–0.65 | 0.04–0.08 | 0.27–0.65 |
| Rawal Lake (54) | Lloyd | Muscles | 1.08–2.15 | 0.09–0.16 | 0.07–0.15 | 0.01–0.12 |
| | | Muscles | | | | |
| Ravi River (14) | Lahore/Baloki | BT | 0.00–2.91 | 0.01–1.34 | 0.01–0.61 | 0.01–0.82 |
| Indus River (185) | Jinnah | Muscle | 0.63–2.34 | 0.09–0.07 | 0.06–0.10 | 0.00–0.73 |
| | Chashma | Muscle | 0.60–3.00 | 0.07–2.33 | 0.02–0.99 | 0.01–1.13 |
| | Taunsa | Muscle | 0.82–2.69 | 0.01–0.65 | 0.05–0.29 | 0.01–0.31 |
| | Guddu | Muscle | 0.82–2.83 | 0.01–0.32 | 0.03–0.62 | 0.01–0.33 |
| | Lloyd | Muscle | 0.21–2.10 | 0.10–0.67 | 0.03–0.21 | 0.01–0.09 |
| Ravi River (5) | Baloki/Sidhani | BT | 75.42–124.90 | | | |
| Fish Hatchery (52) | Islamabad | Muscle | | | | |
| Ravi River (5) | Ravi | Gills/liver | | | 1.10–4.26 | 1.46–6.23 |
| Indus River (5) | Mianwali | Muscles | 251.60–1,179.00 | | | 1.12–10.76 |
| Ravi River (15) | Balloki | Muscles | 34.20–58.90 | 0.84–7.22 | | |
| Indus River (1) | Mianwali | Muscles | 17.80–423.90 | 1.93–577.87 | | 0.13–4.48 |
| Chenab River (8) | Sialkot | Muscles | | 6.14–14.83 | 6.46–13.14 | 31.27–32.77 |
| Sindh Province | | | | | | |
| Indus river (5) | | Muscles | 46.00–80.00 | 4.00–18.40 | 96.20–191.00 | 14.40–23.00 |
| Indus river (63) | | Muscles | 2.95–13.95 | 0.45–1.58 | 0.04–0.05 | |
| Keenjhar Lake (36) | Sunheri | Liver | | | | |
| | Helaya | Gills | | | | |
| | Khumbo | Muscles | | | | |
| Glass aquariums (10) | | BT | | | | |
| Manchar Lake (50) | | BT | 373.40–398.70 | 2.30–9.80 | 1.39–9.30 | |
| Manchar Lake (200) | | Gills | | | | |
| | | Mouth | | | | |
| | | Intestine | | | | |
| Fish hatchery (240) | | Liver | | | | |
| | | BT | | | | 0.46–2.40 |
| Arabian Sea (Marine water fish species) | | | | | | |
| Arabian Sea (30) | | Muscles | | 1.10–10.72 | 0.30–1.03 | 2.12–6.17 |
| Arabian Sea (10) | | Muscles | 4.99–19.83 | 0.83–1.56 | 0.26–0.36 | 5.10–8.51 |
| Arabian sea (143) | | Muscles | 0.14–10.20 | 0.03–0.32 | 0.01–0.15 | 0.03–0.59 |
| Arabian Sea (7) | | Muscles | 0.71–3.41 | 0.12–0.51 | 0.03–0.27 | 0.07–0.11 |
| Arabian sea (145) | | BT | | | | |
| MPL | | | 30.00 | 30.00 | 1.00 | 0.50 |
| MPL | | | 1,000.00 | 30.00 | 2.00 | 5.50 |

Number in parenthesis represents the collected fish samples

[a] Body tissues continued

| Fe | Mn | Ni | Hg | Pb | Ag | As | References |
|---|---|---|---|---|---|---|---|
|  |  | 128.00–133.00 |  | 313.70–321.00 |  |  | Yousafzai et al. (2008) |
|  |  | 97.00–159.30 |  | 682.00–389.30 |  |  | Yousafzai |
|  |  | 122.70–152.00 |  | 453.30–301.30 |  |  | et al. (2010) |
|  |  | 95.30–383.70 |  | 781.70–603.30 |  |  |  |
|  |  | 108.00–111.70 |  | 377.00–623.30 |  |  |  |
|  |  | 106.70–117.70 |  | 528.30–599.30 |  |  |  |
|  |  | 4.70–110.00 |  | 125.70–1,065.30 |  |  | Yousafzai et al. 2012 |
|  | 0.08–0.20 |  |  | 0.04–1.66 |  |  | Khan et al. (2012) |
|  | 0.03–0.09 |  |  |  |  |  |  |
|  | 0.02–0.91 |  |  |  |  |  |  |
| 2.25–6.72 | 0.20–0.29 | 0.52–0.82 | 0.25–1.53 | 1.14–2.46 |  | 0.01–0.11 | Ashraf et al. (1991) |
| 2.17–3.70 | 0.18–0.82 | 0.37–0.80 | 0.77–1.07 | 0.75–1.02 |  | 0.12–0.81 |  |
| 1.55–2.85 | 0.06–0.60 | 0.18–0.34 | 0.58–1.73 | 0.07–0.51 |  | 0.64–1.38 |  |
|  |  |  | 0.24–3.20 |  |  |  | Tariq et al. (1992) |
| 0.01–4.21 | 0.01–0.73 | 0.01–2.21 | 0.01–3.40 | 0.01–3.40 | 0.01–2.17 | 0.01–1.31 | Tariq et al. (1994) |
| 0.47–1.82 | 0.03–0.99 | 0.05–0.69 | 0.08–1.82 | 0.06–1.99 | 0.87–0.01 | 0.32–2.70 | Tariq et al. (1995) |
| 0.18–6.10 | 0.01–0.92 | 0.08–0.80 | 0.10–0.35 | 0.01–2.35 | 0.01–2.18 | 0.05–2.80 |  |
| 1.20–2.40 | 0.01–1.11 | 0.01–0.23 | 0.06–3.01 | 0.03–0.70 | 0.82–2.69 | 0.06–3.01 |  |
| 0.92–6.31 | 0.01–0.47 | 0.01–0.52 | 0.09–3.92 | 0.01–0.59 | 0.02–1.09 | 0.01–3.07 |  |
| 0.88–2.17 | 0.12–3.05 | 0.03–0.17 | 0.10–1.63 | 0.02–0.91 | 0.08–0.93 | 0.47–1.07 |  |
| 291.40–431.30 | 9.02–17.49 | 1.73–9.45 |  | 7.58–12.55 |  |  | Javid (2005) |
| 133.60–538.00 | 2.50–8.80 |  |  |  |  |  | Ansari et al. (2006) |
|  |  |  |  |  |  |  | Rauf et al. (2009) |
|  | 1.65–15.17 |  | 1.76–8.72 | 0.61–3.84 |  |  | Jabeen and Chaudhry (2010) |
|  |  |  | 28.40–126.30 | 1.90–4.40 |  | 39.00–66.50 | Nawaz et al. (2010) |
|  | 0.18–62.28 |  | 0.22–4.11 | 0.21–5.28 |  |  | Chaudhry and Jabeen (2011) |
|  |  |  |  | 4.23–30.06 |  |  | Qadir and Malik (2011) |
|  |  |  |  | 0.60–50.60 |  | 1.40–6.00 | Gachal et al. (2006) |
| 7.10–33.73 | 0.42–7.31 | 0.31–0.76 |  | 0.46–1.49 |  |  | Dewani et al. (2003) |
|  |  |  |  | 0.70–2.39 |  |  | Korai et al. (2008) |
|  |  |  |  | 0.89–2.68 |  |  |  |
|  |  |  |  | 0.74–2.25 |  |  |  |
|  |  |  |  | 34.90–41.37 |  |  | Javid et al. (2007) |
| 1,517.90–6,670.30 |  | 2.20–4.60 |  | 2.40–11.60 |  | 2.35–7.50 | Arian et al. (2008) |
|  |  |  |  |  |  | 1.01–11.20 | Shah et al. (2009) |
|  |  |  |  |  |  | 1.01–18.60 |  |
|  |  |  |  |  |  | 1.01–11.22 |  |
|  |  |  |  |  |  | 3.51–10.91 | Ahmad and Bibi (2010) |
|  |  |  |  | 4.21–4.92 |  |  |  |
| 910.00–1,089.00 | 2.92–10.70 | 0.80–3.25 | 0.01–0.63 | 0.89–4.30 | 0.10–0.99 | 0.39–2.10 | Tariq et al. (1991a, b) |
| 889.00–1,791.00 | 4.87–7.68 | 12.09–18.28 | 0.09–0.16 | 0.14–11.63 | 0.29–0.53 |  | Tariq et al. (1993) |
| 0.31–5.91 | 0.03–0.19 | 0.04–1.89 | 0.01–0.75 | 0.01–0.66 | 0.02–0.56 | 0.14–7.32 | Jaffar et al. (1995) |
| 1.14–3.59 | 0.06–0.20 | 0.15–0.31 | 0.10–0.41 | 0.02–0.28 | 0.04–0.27 | 0.35–1.96 | Tariq et al. (1998) |
|  |  |  | 0.73–1.47 |  |  |  | Shah et al. (2010) |
| 100.00 | 1.00 | 1.00 | 0.10 | 0.30 | – | 0.50 | FAO/WHO (1983) |
|  |  | 5.50 | 0.50 | 2.00 |  | 2.00 | ANMHRC/WAFDR Rahman et al. 2012 |

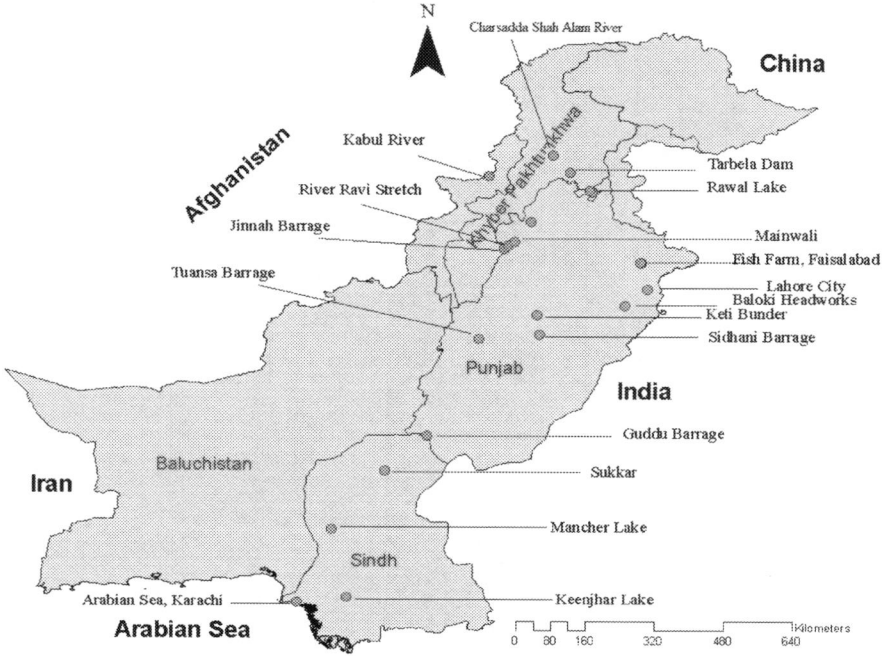

**Fig. 1** The location of study sites in Pakistan at which fish were harvested for HM analyses as addressed in the body of this paper

### 2.1.1 Khyber Pakhtunkhwa Province

The River Kabul is a main source of water in the Khyber Pakhtunkhwa. The River Kabul originates from Afghanistan and passes through the most populous areas and industrial cities of Peshawar, Charsadda, and Nowshera, and then joins the Indus River. This river provides a means of livelihood to thousands of people living along its banks and tributaries. Yousafzai et al. (2008) analyzed gill tissue for residues of Cu, Ni, Pb, Cr, and Zn in the fish species *Tor putitora*. Samples were collected from the Kabul River at two different sites: Nowshera District (contaminated site) and Warsak Dam (background site). Results have shown that the highest concentration in fish was for Zn (1,224.90–2,414.00 µg/g), whereas the lowest rate was for Cr (6.02–6.60 µg/g) in the study area (Table 1). The descending order of the concentration discovered by these authors in fish was as follows: $Zn > Pb > Ni > Cu > Cr$. The authors concluded that fish at the contaminated site absorbed higher levels of HM than occurred at a control site, where background levels existed.

Two years later, Yousafzai et al. (2010) conducted a comprehensive study on the Kabul River near the Nowshera district to determine the level of HM residues in two fish species (*Wallago attu* and *Labeo dyocheilus*) living in different feeding zones of

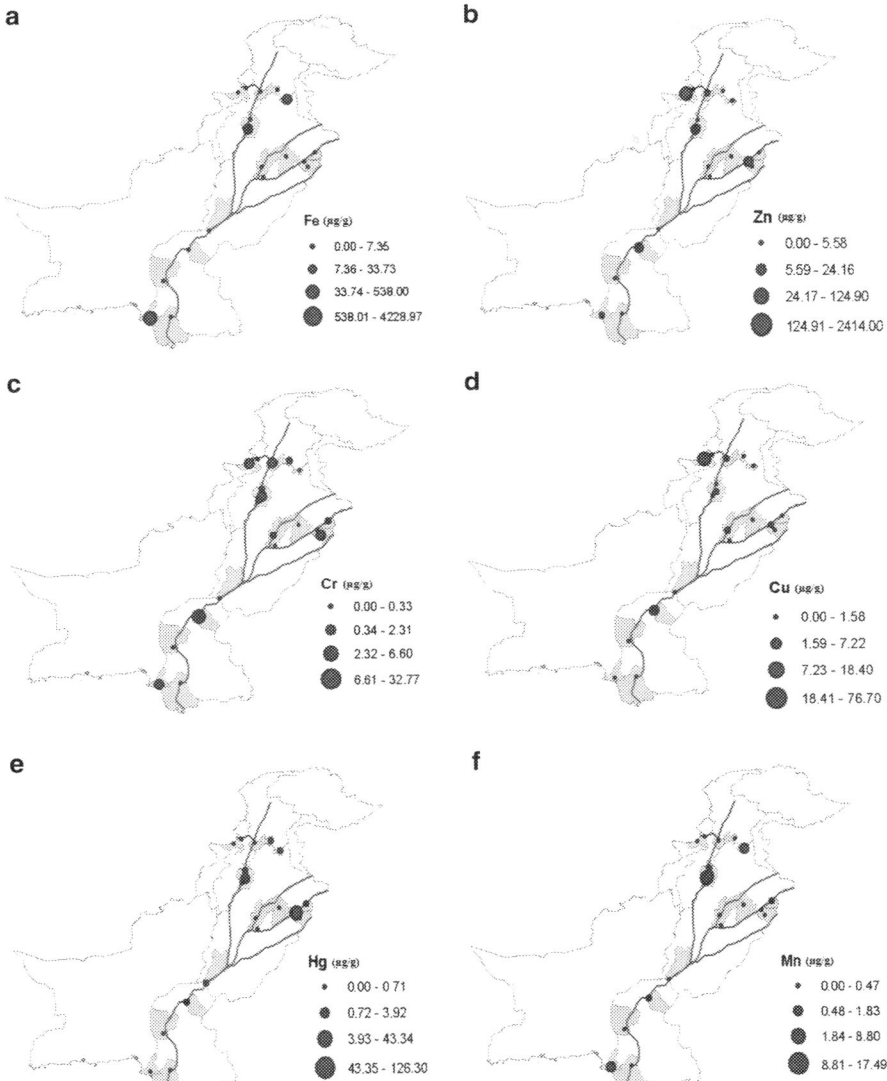

Fig. 2 A depiction of the spatial distribution of different HM concentrations found in fish at the sampling sites shown on the maps

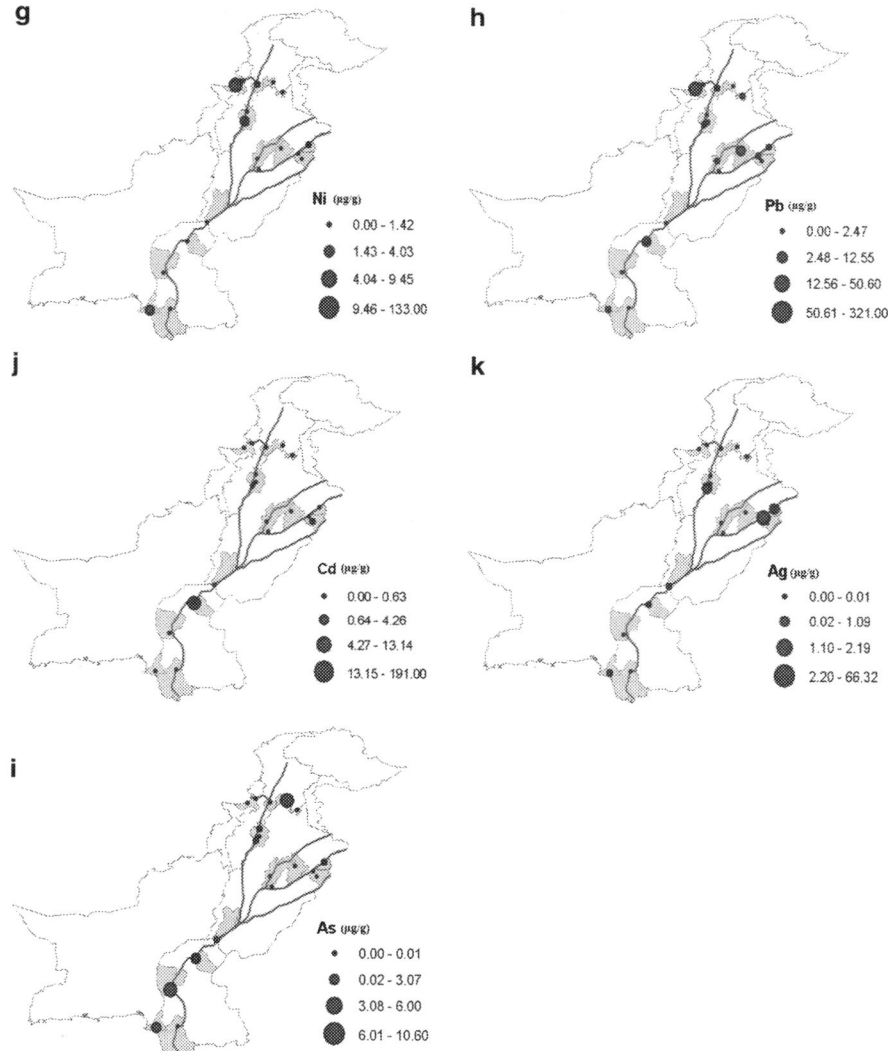

**Fig. 2** (continued)

the same habitat. In the study area, the concentrations were highest for Zn (1,971 µg/g) and lowest for Cd (74 µg/g) (Table 1). The descending order at the studied site for HM residues was as follows: Zn>Cr>Cu>Pb>Ni>Cd. Similarly, the order of HM concentrations in different organs of *Wallago attu* was reported to be: skin>gills>muscles>intestine>liver, whereas that of *Labeo dyocheilus* was liver>muscles>skin>intestine>gills. The authors concluded that *Labeo dyocheilus* took up 65.2% more toxic HM residues than did *Wallago attu*. This result indicated that *Labeo dyocheilus*, which is an omnivorous fish, may take more HM up from natural habitats than does *Wallago attu*, which is a carnivorous fish.

Recently, a comprehensive study has been conducted by Khan et al. (2012) to investigate the occurrence of Pb, Cu, Zn, and Mn concentrations in the Shah Alam River (River Kabul) fish species. Results revealed that *Cyprinus caprio* exhibited the highest concentrations of Zn in the liver, gills, and muscles, whereas Cu and Mn had lower levels. However, Pb concentrations (0.04 µg/g) were observed only in liver. HM uptake by *Cyprinus caprio* occurred at levels in the descending order of gills > liver > muscles. The highest Cu and Zn concentrations were determined in liver and gills of *Cirrhinus marigala*, with Mn levels being lower. The highest Mn levels were found in muscles. Liver exhibited a lower Pb concentration (0.17 µg/g). The highest concentrations found in *Mystus bleekeri* were of Cu, followed by levels of Zn and Mn. In the same species, the highest Pb concentration was found in liver (1.17 µg/g). The general pattern of HM residue was in the order of gills > liver > muscles.

In another study, Yousafzai et al. (2012) reported the HM (Zn, Ni, Cr, Cu, and Pb) residue contents in body tissues of common crap (*Cyprinus carpio*) sampled from the Kabul River of the Nowshera district. Results showed that HM residues in fish were highest (826.00–8,391.30 µg/g) for Zn, and lowest (53.30–66.70 µg/g) for Cd (Table 1). There was a descending pattern of HM residues in fish tissues as follows: intestine > skin > liver > gills > muscle. The intestine exhibited the highest HM residues, which resulted from direct dietary exposure; muscle tissue levels of HM were minimal.

River Kabul fish samples collected from sites within the Nowshera district showed multifold higher concentrations than those at the Warsak and Peshawar sites. These multifold concentrations in Nowshera fish may have resulted from several sources that include agricultural runoff, pollution from surrounding mining operations, or releases of urban and untreated industrial sewage from the nearby Amangarh Industrial Estate and Peshawar city (Yousafzai et al. 2008; Khan et al. 2011, 2012). Concentrations of HM (e.g., Zn, Cu, Cd, Cr, Ni, and Pb) in fish collected from the River Kabul were found to exceed the maximum permissible limits (MPL) established by FAO/WHO (1983) and The Australian National Health and Medical Research Council (ANMHRC)/Western Australian Food and Drug Regulation (WAFDR) as reported by Plaskett and Potter (1979) and Rahman et al. (2012) (Table 1).

### 2.1.2 Punjab Province

Punjab hosts several rivers (viz., Indus, Chenab, Sutlej, Ravi, and Jhelum), which form the upper Indus plain, and remain one of the most agriculturally rich and fertile of areas. These rivers provide water for agricultural and domestic uses (Khan et al. 2013). Ashraf et al. (1991) reported HM concentrations in three fish species, viz., *Rita rita*, *Wallago attu*, and *Cirrhinus mrigala* from three sites that included the Tarbela reservoir, Chashma, and Lloyd barrage, along the Indus River. Among the fish species studied, *Rita rita* collected from the Tarbela reservoir showed the highest (6.72 μg/g) concentrations for Fe. The estimated pattern of HM residues in fish species from these sites were found in the order *Rita rita* > *Wallago attu* > *Cirrhinus mrigala*.

One year later, Tariq et al. (1992) reported Hg concentrations (1.16–1.38 μg/g) in two fish species (viz., *Catla catla* and *Chela chanius*) collected from the Rawal Lake, Islamabad/Rawalpindi. The uptake of Hg in male *Chela chanius* was 0.07 μg/g per unit length, whereas the female *Chela chanius* displayed a smaller uptake rate of Hg (0.04 μg/g) per unit length. Moreover, the Hg concentration in male *Catla catla* was twice that of the females. This study also revealed species specificity regarding Hg uptake in both fish species studied. The maximum Hg residue was documented in male specimens, which is in line with the results reported by Lyle (1984) and Marcovecchio et al. (1986).

Tariq et al. (1994) sampled 14 commercial fish species from the Ravi River at Lahore and Baloki sites and analyzed the collected specimens for HM (viz., Ni, Ag, Cu, Cr, and Mn, Cd, As, Hg, Pb, Zn, and Fe) contents. Results showed that the highest HM concentrations were found in *Labeo calbesu* (Ag 2.17 μg/g), *Mystus vitattus* (Cd 0.62 μg/g), *Labeo rohita* (Mn 0.73 μg/g and Pb 3.40 μg/g), and *Labeo gonius* (Cu 1.18 μg/g) collected from the Lahore site. Similarly, the fish species *Channa marulius* (Ag 2.17 μg/g), *Labeo gonius* (Cd 0.62 μg/g), *Catla catla* (Cu 1.34 μg/g) showed maximum concentrations of HM at the Baloki site. The study also demonstrated that *Labeo gonius* and *Labeo rohita* could serve as indicators of HM pollution, because they showed their maximum concentrations at both the Lahore and Baloki sites.

Tariq et al. (1996) also investigated the presence of HM in various fish species captured at the Chashma, Jinnah, Guddu, Lloyd, and Taunsa barrages along the Indus River. The highest concentrations of Hg (3.92 μg/g) and As (3.07 μg/g) were found in fish muscle tissue from the Guddu barrage (Table 1). Among fish species evaluated, specimens of *Hetroptirus fossilus* had the highest concentrations of Ag (2.19 μg/g), Cu (2.34 μg/g), and Pb (1.99 μg/g) in muscles. Similarly, *Labeo calbasu* showed the highest concentrations of As (3.07 μg/g) in specimens taken from Guddu barrage. *Mystus vitatus* samples collected from the Chashma barrage revealed high concentrations of Cd (0.99 μg/g) and Fe (6.10 μg/g). *Tilapia nilotica* showed the highest concentration of Mn (3.05 μg/g) when test specimens were collected from the Lloyd barrage. The trend of HM concentrations found in the muscle tissue of the different fish species tested was as follows: Ag > As > Cu > Fe > Cd > Cr.

Javid (2005) reported HM concentrations in body tissues of *Cirrhina mrigala*, *Labeo rohita*, and *Catla catla* specimens collected from the Ravi River (Baloki headworks to Sidhani barrage) and its tributaries. The authors reported the highest (291.40–431.30 μg/g) concentrations for Fe, whereas the lowest (1.73–9.45 μg/g) levels were for Ni at the Baloki site (Table 1). The highest concentrations of HM were found in fish taken from the Baloki site. The authors reported that fish muscles, kidney, and liver were the main depository organs for HM. The magnitude of HM among sampled fish species was in the following order: *Catla catla* > *Labeo rohita* > *Cirrhina mrigala*, whereas the order of the concentration in tissues was liver > kidney > gills > muscles.

Ansari et al. (2006) studied wild fish (*Puntius chola*) for the magnitude of Zn, Fe, Cu, Cd, and Mn residues. Results showed that, except for Cu and Cd, the concentrations of HM taken up were directly related to the increase in body weight of *Puntius chola*. However, Zn, Fe, Cu exhibited negative allometry to an increase in HM concentrations, because the residue levels were less than the rate of excretion as the fish grew.

Rauf et al. (2009) performed an interesting study on the residues of Cd and Cr in *Cirrhina mrigala*, *Labeo rohita*, and *Catla catla* that were collected from the Lahore Siphon, Shahdera Bridge, and Baloki headworks in the Ravi River, Pakistan. The concentration of HM found in these fish varied considerably, and depended upon the fish species, tissue examined and collection site. Results showed that liver had the highest level of Cd (4.26 µg/g) and Cr (6.23 µg/g), whereas gills had the lowest concentrations of Cd (1.10 µg/g) and Cr (1.46 µg/g) (Table 1). *Catla catla* exhibited the highest concentration of Cd (2.58 µg/g) and Cr (3.58 µg/g). The authors concluded that the Balloki headworks had the highest contamination level, and this level may be attributed to industrial effluents, agricultural runoff, and domestic sewage releases.

Jabeen and Chaudhry (2010) reported HM residues in common carp (*Cyprinus caprio*) and *Oreochromis mossambicus* at the Chashma and Shebhaz Khel in the Mianwali district, along the Indus River. Among the HM detected, Zn (251.60–1,179.00 µg/g) displayed the highest concentrations in fish; the lowest reported concentrations were for Hg (1.76–8.72 µg/g) (Table 1). Maximum Zn, Pb, and Mn concentrations were found in fish tissues collected at the Chashma site, whereas Cr and Hg levels were highest at the Shebhaz Khel. The pattern of HM residues among tissues varied greatly, when specimens were collected from the Chashma verses Shabazkhel.

Nawaz et al. (2010) reported Pb, Hg, As, Cu, and Zn residues in either edible or nonedible fish species collected from two sites of the River Ravi at the Balloki headworks. The highest As concentration (66.50 µg/g) was found in *Cirrhinus mrigala* (edible fish), whereas the lowest (39.00 µg/g) was in a *Mystus vittatus* specimen (nonedible fish). Water of the Lahore and Kasur districts had the highest (32–47 µg/L) As concentrations (Farooqi et al. 2007). Maximum (126.30 µg/g) Hg concentrations were found in *Notopterus notopterus* (edible fish) and minimum (28.40 µg/g) ones in *Xenentodon cancila* (nonedible fish). The Hg levels were high in both edible and nonedible fish species. However, the source of Hg contamination that caused these residues is still unknown and needs further study. Edible fish species such as *Labeo rohita* showed the highest (58.90 µg/g) concentrations of Zn, while the lowest (34.20 µg/g) levels were in *Cirrhinus mrigala*. Similarly, Cu and Pb concentrations in both edible and nonedible fish ranged from 1.93 to 577.87 µg/g and 1.90 to 4.40 µg/g, respectively.

Chaudhry and Jabeen (2011) reported the effects of HM contamination in organs of *Labeo rohita*, collected from the Indus River in the Mianwali district. The order in which HM concentrations appeared in fish tissues was as follows: liver>gills>skin>muscles at the Shehbaz Khel and Chashma collections sites, whereas at Kukranwala the order was: gills>liver>skin>muscles. Mn, Cr, Pb, and Zn showed maximum concentrations in gills, whereas Hg and Cu exhibited maxima in the liver. Results revealed that gills were the most susceptible organ to HM contamination, followed by liver, skin, and fish muscle tissue. The study authors concluded that high HM concentrations exist in fish organs from specimens collected at the Shehbaz Khel and Chashma.

Qadir and Malik (2011) reported that four HM (viz., Pb, Cu, Cr, and Cd) were concentrated in the organs (viz., gills, kidneys, liver, and muscles) of eight fish species. Fish in this study were sampled from two tributaries (Nullah Palkhu and Nullah Aik) of the River Chenab, Pakistan during both post and pre-monsoon seasons. Pb and Cr residues in fish collected during the pre-monsoon period existed in the following descending order: gills>liver>kidneys>muscles, whereas Cd and Cu residue levels showed liver>gills>kidneys>muscles. The authors reported the highest mean concentrations of Pb in gills (30.06 µg/g) and muscles (9.53 µg/g) of *Heteropneustes fossilis*, and the lowest levels in gills (15.72 µg/g) of *Wallago attu*, or in muscle tissues (4.23 µg/g) of *Cirrhinus punctata* sampled from the Nullah Aik. The maximum mean concentration of Cd was reported in liver (13.14 µg/g) of *Mystus cavasius* (post-monsoon season), whereas the minimum occurred in the gills (6.46 µg/g) of *Cirrhinus punctata* (pre-monsoon season) sampled from the Nullah Palkhu. The highest mean concentration of Cr in gills (31.27 µg/g) was detected in a *Wallago attu* specimen collected from the Nullah Aik (post-monsoon period). Similarly, the maximum average concentration of Cr was reported in gills (32.77 µg/g) from both Nullah sites (pre-monsoon period). The maximum mean concentration of Cr was found in *Heteropneustes fossilis*, whereas the minimum was detected in a specimen of *Puntius sophore*. The highest mean concentration of Cu was found in the liver (14.83 µg/g) of a *Puntius sophore* specimen sampled from the Nullah Aik (post-monsoon period). Similarly, the maximum concentration of Cu was recorded in liver (6.14 µg/g) of a *Cirrhinus punctata* specimen sampled from the Nullah Aik (pre-monsoon period). Results showed that *Puntius sophore* and *Labeo rohita* accumulated the highest mean concentrations of Cu, whereas the lowest Cu levels were in a *Wallago attu* sample.

The River Indus and its tributary waters are highly contaminated with HM (Yousafzai et al. 2008; Muhammad et al. 2011a; Khan et al. 2013). The HM contamination of such waterways results from release of household and urban sewage, agricultural runoff, mobilization from bedrock, or from mining operations or release of untreated industrial waste in the catchment area (Yousafzai et al. 2008; Muhammad et al. 2010, 2011b; Khan et al. 2013). As a result, these HM are taken up in fish species that then display concentrations that greatly exceed the MPL values set by regulating entities, such as the FAO/WHO (1983) and ANMHRC/WAFDR (Rahman et al. 2012) for the respective HM contaminants (Table 1). The degree to which HM concentrate in fish greatly varied from species to species and by location as reported by Jabeen and Chaudhry (2010). Fish sampled from streams and rivers located along or near urban areas or industrial estates/complexes revealed a higher concentration than others lacking such proximity to sources of potential HM release (Tariq et al. 1995; Javid 2005; Rauf et al. 2009; Chaudhry and Jabeen 2011) (Fig. 2). The levels of HM that occur in individual fish species depends on the feeding source and position in the food chain of the consuming fish (Tariq et al. 1993; Yousafzai et al. 2008, 2010). Among fish tissues, liver, kidney, gill, skin, and muscles were the primary depository organs (Javid 2005; Rauf et al. 2009; Chaudhry and Jabeen 2011). Of main concern to humans, is the level of HM residues that ends up in muscle tissue. Humans consume the HM-contaminated muscle tissue, and at certain levels these contaminants may pose a potential threat to health (Khan et al. 2012; Yousafzai et al. 2012).

### 2.1.3 Sindh Province

Ahmad et al. (2004) reported that 16–36% of the population of Sindh province has been exposed to high levels (10–50 ppb) of As-contaminated water. Humans that consume excessive As may suffer toxic effects to liver, bladder, and skin (cancer). Tariq et al. (1996) and Wallace et al. (1977) reported that release of domestic sewage, agricultural runoff, and untreated industrial effluents find their way into the Indus River at numerous places, and HM may exist in these releases at levels that impair water quality.

Dewani et al. (2003) investigated the HM contamination in fish species collected from the polluted Fuleli canal. The fish species studied revealed great variability in the amount of HM retained. In all fishes examined, Fe was the most prominent residue, and Cd was the lowest. The highest Fe and Zn concentrations (33.73 and 13.95 µg/g) were found in *Eutropiicheythys vacha*, while the lowest concentration was for Fe (7.10 µg/g), which was detected in a specimen of *Wallago attu*. The highest Mn level (7.31 µg/g) was recorded in the muscles of *Rita rita*, whereas *Wallago attu* exhibited the lowest Mn concentration (0.42 µg/g). The highest Cu concentration (1.58 µg/g) was detected in samples of *Labeo dero*. High concentrations of Pb (1.49 µg/g), Ni (1.76 µg/g) and Co (0.41 µg/g) were also documented to occur in the sampled fish species.

Scientists have proposed that the Indus Dolphin be considered as an endangered species across the globe, because it faces risks from many water pollutants (Gachal et al. 2006). Gachal et al. (2006) reported that Cd displayed the highest level of any HM in the Indus Dolphin (96.20–191.00), whereas As levels were the lowest (1.40–6.00 µg/g). Other HM (viz., Zn, Cu, Cr, and Pb) displayed residue levels between these two extremes (Table 1).

In addition, Javid et al. (2007) reported the presence of Pb in *Labeo rohita*, *Cirrhina mrigala*, and *Catla catla* collected from glass aquaria at room temperature. *Catla catla* was observed to be the most sensitive to Pb pollution, followed by *Labeo rohita* and *Cirrhina mrigala*. The Pb concentration was highest in *Labeo rohita* (41.37 µg/g), followed by *Cirrhina mrigala* (35.12 µg/g) and then *Catla catla* (34.90 µg/g).

Korai et al. (2008) investigated the presence of Pb concentrations in organs of *Catla catla* captured from three sites (Sunheri, Helaya, and Khumbo) of the Keenjhar Lake over a 3-year period. The Pb concentration present in the tissues of *Catla catla* varied with tissue, with the highest (0.89–2.68 µg/g) level in liver, and the lowest level (0.74–2.25 µg/g) in muscles during January 2003 to December 2005. The concentration of Pb (2.19 µg/g) was observed to be highest in fish liver sampled from the Helaya site, whereas the lowest (1.14 µg/g) was recorded in muscle tissue from fish collected from the Khumbo site during 2003. Moreover, the Pb concentration varied between 1.14 and 2.21 µg/g in fish tissues collected during 2004. The highest Pb concentration (2.21 µg/g) was detected in liver at the Helaya site, where as the lowest (1.14 µg/g) was in muscle tissue from fish at the Khumbo site. Liver showed the highest (2.68 µg/g) concentration of Pb in fish captured from the Sunheri site, whereas muscles revealed the lowest concentration (0.59 µg/g) in fish sampled from the Helaya site. The authors concluded that liver and gills had a higher affinity for concentrating Pb than did muscle tissue. The authors suggested that the main sources of Pb contamination in the Keenjhar Lake were contaminated wastewater released from households, and industry and runoff from agriculture land located in the surrounding areas.

Arian et al. (2008) investigated the HM concentrations in tissues of fish specimens (*Oreochromis mossambicus*) collected from the Manchar Lake. Liver showed the highest (6,670.30 µg/g) concentration for Fe, where as the lowest (0.46 µg/g) were recorded for Cr (Table 1). The descending order of HM concentrations in body tissues was as follows: Fe>Zn>Pb>As>Cu>Ni>Cd>Cr, whereas that of liver was Fe>Zn>Cu>Cd>Pb>As>Ni>Cr. HM concentrations in tissues greatly varied, which may be attributed to differences in their adsorption and retention capacity. The authors concluded that the fish species collected from the Manchar Lake had higher As concentrations than did those of the Indus River. The most important sources of HM contamination in the Manchar Lake were from the Nara Valley drainage that carries releases from ore mining, dye manufacturing and tannery activities, household and industrial wastewater releases, and runoff of pesticides from agricultural land (Sarkar and Datta 2004).

Ahmad and Bibi (2010) evaluated the uptake of Pb concentrations in the tissues of *Catla catla*, sampled from polluted and control sites. The Pb concentration was highest (4.92 µg/g) in skin, and lowest (4.21 µg/g) in intestine of fish collected from the polluted site. Similar Pb concentrations from fish at the polluted site existed in internal fish tissues, as follows: liver (4.79 µg/g), muscles (4.41 µg/g), and intestine (4.21 µg/g). The authors concluded that the source of Pb concentration in fish tissues (viz., skin, liver, gills, eyes, muscles, and intestine) of the study area may be attributed to untreated industrial effluents and municipal wastewater releases. In addition, the main sources of Pb contamination in the aquatic environment were from the fertilizer industry, planting processes, ore refining and burning of Pb-containing gasoline that leaked from fishery boats, and municipal and industrial sewages (Handy 1994).

## 2.2 Marine Fish

Toxic HM contamination, which is persistent and bioaccumulative, increasingly threatens marine ecosystems (Balkas et al. 1982). Karachi, Pakistan is a coastal city facing multiple urbanization and industrialization problems. Therefore, nearby coastal areas of the Arabian Sea are receiving a huge quantity of unregulated industrial sewage releases that ultimately affect aquatic life (Jaffar et al. 1995; Tariq et al. 1998).

Tariq et al. (1991a, b) performed a very important study, in which HM residues were investigated in fish species that were sampled along coastal areas near Karachi. Ten fish species were collected from near-shore (5–10 km) and off-shore (15–40 km) sites. Results revealed that Hg and Fe were highly concentrated in *Loligo duvauceli* and Cr, Pb, Mn, and Ni in *Sardinella longiceps*, whereas Cd was present in *Lapturacanthus savala*. As mentioned earlier, fish can serve as bioindicators of HM pollution of marine ecosystems, particularly in coastal waters (Tariq et al. 1991a, b; Jaffar et al. 1995). The authors of this study (Tariq et al. 1991a, b) concluded that two species could be used to effectively biomonitor for HM pollution along the coast of the Arabian Sea. *Sardinella longiceps* is suited for tracing the extent of HM contamination for Cr, Pb, Mn, and Ni, whereas *Loligo duvauceli* is most suited to trace contamination by Hg and Fe.

Tariq et al. (1993) reported the presence of HM in fish sampled off-shore along the Arabian Sea coast. They sampled coastal fish species (*Rastrelliger kanagurta* and *Pomadasys maculatus*), that primarily feed on invertebrates and small fish. However, *Chaetodon jayakeri*, which is a migratory coastal fish that feeds on invertebrates and small fish, was also sampled (Bianchi 1985). The highest residues recorded from this experiment of any HM for all species were for Fe (889.00–1,791.00 µg/g), whereas the lowest residues (0.09–0.16 µg/g) were recorded for Hg. However, other HM were present in the fish at levels that were between these two extremes (Table 1). The highest HM residues in *Rastrelliger kanagurta* were for Fe (1,791 µg/g). The lowest HM residues were for Hg (0.09 µg/g), which appeared in specimens of *Chaetodon jayakeri*. The authors concluded that HM residues vary considerably among fish species and their uptake is species specific.

Jaffar et al. (1995) investigated HM concentrations in fish species collected along the southwest coast of the Arabian Sea. The highest HM values found were for Zn (0.14–10.20 µg/g), and the lowest (0.05–0.15 µg/g) were for Cd (Table 1). The authors concluded that the mean concentration of toxic HM in fish species were higher for those that had been discharged to waters from industrial activities (viz., Fe, As, Cu, Hg, Cr, Zn, Cd, and Ni).

Tariq et al. (1998) reported toxic HM residue levels in muscle of seven marine fish species that were sampled from the southwest coastal area of the Arabian Sea. Among toxic HM, Zn and Fe showed the highest (3.41 µg/g and 3.59 µg/g) concentrations, respectively. *Argyrops spinifer* showed the highest (0.27 µg/g) concentration for Ag, whereas *Chaetodon jayakeri* had the lowest (0.04 µg/g). Similarly, *Atrobucca trewavasii* revealed the highest (1.96 µg/g) concentration for As, whereas *Acanthopagarus lactus* had the lowest (0.35 µg/g) level. Moreover, *Argyrops spinifer* showed the highest levels (viz., 3.59 µg/g and 0.20 µg/g) for Fe and Mn, respectively.

Shah et al. (2010) studied the effects of total Hg concentration in muscle tissues of four marine fish species. The Hg concentrations measured in muscles ranged from 0.71 to 1.41 µg/g. At this level, daily intake of Hg residues in 250 g of fresh fish muscle would exceed the WHO permissible human consumption limit (0.22 µg/person/day) (WHO 1989).

Tabinda et al. (2010) studied HM residues in six fish species collected from the coastal waters at the Keti Bunder, Thatta, of Sindh province. The highest (3.60 µg/g) concentrations were measured for Pb, whereas the lowest (0.01 µg/g) were for As. These results indicated that the Pb and Cu concentrations appearing in *Pampus argenetus* and *Tenualosa ilisha*, and the Cr levels in most remaining fish species, exceeded the recommended limits of the FAO/WHO (1983) and ANMHRC/WAFDR (Rahman et al. 2012) as shown in the Table 1.

## 2.3 Heavy Metal Residues in Fish Species of Neighboring Countries

Although the focus of this review is HM fish residues in Pakistan, we realize that neither Pakistan nor the aquatic species that exists in and around it are wholly isolated from adjoining geographical areas. Therefore, we thought that comparing selected data on fish residues from neighboring countries to those in Pakistan would be instructive and useful to readers of our review.

### 2.3.1 Bangladesh

Amin et al. (2011) reported the Zn (3.14–186.90 µg/g), Mn (4.10–51.67 µg/g), Cu (1.48–21.30 µg/g), Ni (1.80–8.40 µg/g), Pb (0.5–4.05 µg/g) residue levels that existed in muscle tissue of fish captured from the Gumti River (Bangladesh). Similarly, Rahman et al. (2012) reported the HM (Pb, Cd, Cu, Cr, Ni, Zn, As, and Zn) concentrations in fish sampled from the Bangshi River. The concentrations of HM in fish were as follows: Zn (42.83–418.00 µg/g), Mn (9.43–51.17 µg/g), Cu (8.33–43.18 µg/g), Pb (1.76–10.27 µg/g), As (1.97–6.24 µg/g), Ni (0.69–4.36 µg/g), Cr (0.47–2.07 µg/g), Cd (0.09–0.87 µg/g). These concentrations of HM (Zn, Ni, Cr, and Cd) were within the MPL set by ANMHRC/WAFDR for human consumption. HM residues in fish species of Bangladesh were found to be lower than those reported by Yousafzai et al. (2008, 2010, 2012) for fish in Pakistan.

### 2.3.2 China

Qiu et al. (2011) studied the mean concentrations of Pb, Cu, Zn, Cd, Cr, Hg, and As in poppano and snapper sampled from Daya and Hailing Bays. The resulting respective residue levels were: 2.7, 1.6, 27.3, 0.025, 0.62, 0.18, and 0.59 µg/g in pompano, and 2.6, 1.5, 23.6, 0.020, 0.55, 0.22, and 0.53 µg/g in snapper. Similarly, Fu et al. (2013) analyzed the residue levels of HM such as Zn (54.09–367.39 µg/g), Cu (1.06–83.88 µg/g), Pb (4.14–27.18 µg/g), Cr (1.01–6.11 µg/g), and Cd (0.31–1.76 µg/g) in fish species collected from the Yangtze River and Taihu Lake, Jiangsu Province. The concentrations of various HM (Cu, Zn, Cd, Cr, Hg, and As), except for Pb were within the MPL set by ANMHRC/WAFDR for human consumption.

HM residue levels in Chinese fish species were found to be lower than those reported by Javid (2005) and Qadir and Malik (2011) for Pakistan.

### 2.3.3 India

Javed and Usmani (2011) reported HM residues in fish species collected from the Rasagani, a popular fish market in Aligarh. Results revealed that the highest (39.00–1,850.00 µg/g) HM levels were for Fe, Cu (9.00–1,250.00 µg/g) and Zn (42.00–459.40 µg/g), followed by Ni (10.80–187.50 µg/g), Mn (1.00–109.40 µg/g), Cr (1.00–27.00 µg/g), and Co (3.00–25.00 µg/g). Similarly, Gummadavelli et al. (2013) investigated HM residue levels in the muscles of *Cyprinus carpio communis* collected from Edulabad Water Reservoir (EBWR) in Andhra Pradesh. The concentrations found in fish muscle tissues were: Pb (290.00–702.00 µg/g), Fe (399.00–1,232.00 µg/g), Ni (236.00–464.00 µg/g), Cr (461.00–798.00 µg/g), and Cd (333.00–883.00 µg/g). The authors of these studies concluded that all HM except Zn have multifold higher concentrations than the MPL set by the FAO/WHO and ANMHRC/WAFDR. These study results further suggested that the HM residue levels found in these fish species render them not fit for human consumption, and consuming them would pose human health risks. We conclude that HM residue levels in Indian fish species were higher than those in Pakistan as reported by Chaudhry and Jabeen (2011) and Khan et al. (2012).

### 2.3.4 Iran

Ebrahimi and Taherianfard (2010) reported the concentration of HM (Cd, Pb, Hg, and As) in fish species caught from three sites of the Kor River. The highest mean concentration reported in fish for different HM were: 0.11 µg/g for Cd, 1.84 µg/g for Pb, 1.14 µg/g for Hg, and 0.98 µg/g for As. Alhashemi et al. (2012) investigated the HM (Cd, Pb, Mn, Co, Ni, Cr, Zn, and Cu) concentration in fish species collected from wetland in the southwest of Iran. The HM concentration ranges reported were: Zn (28.57–49.50 µg/g), Mn (1.03–24.80 µg/g), V (4.87–6.75 µg/g), Pb (2.90–6.39 µg/g), Cu (3.21–5.00 µg/g), Ni (0.96–2.63 µg/g), Cr (0.7–2.60 µg/g), Co (0.41–1.26 µg/g), and Cd (0.13–0.41 µg/g). The HM (Hg, Mn, Pb) residue levels exceeded the MPL set by ANMHRC/WAFDR. HM concentrations in Iranian fish species were found to be lower than those reported by Arian et al. (2008) and Nawaz et al. (2010) for fish in Pakistan.

Generally, the HM concentrations in fish collected from neighboring countries (viz., Bangladesh, China, and Iran) were lower than residue levels found in Pakistani fish. However, fish collected in India were an exception, because their HM levels tended to be higher than those from Pakistan. Yousafzai et al. (2010, 2012), Javed and Usmani (2011) and Gummadavelli et al. (2013) reported that the higher HM residue levels in Pakistani and Indian fish may be attributed to disposal of untreated domestic and industrial waste into the waterways and tributaries of these two countries.

## 3 Conclusions and Recommendations

In general, samples of fish collected from both fresh and marine water sources in Pakistan exhibited HM residues. Some of the more common HM found in fish were As, Fe, Zn, Pb, Cd, Hg, Ni, and Cu. Fish collected from contaminated sites (near urban and industrial estates) showed the highest concentrations of HM. This reflected the reality that wide scale discharge of HM to water occurs from releases to household and urban sewage, from industrial effluents and from disposal of solid wastes to waterways. We expect that HM also contaminate environmental matrices in other urban and industrial areas of Pakistan, although the data to confirm this are still unavailable. We believe there is urgency in performing monitoring studies that are more comprehensive than those that now exist, so as to define the status, distribution, and sources of these HM in the environmental matrices of Pakistan.

Our key conclusions and recommendations are as follows:

1. Although few studies have been conducted in Pakistan on the degree of HM contamination of aquatic species, the data available discloses that residues of HM are present in fish and dolphins, sometimes at excessive levels that may pose a danger to both the exposed species and to humans who consume them for food. The major sources of HM in the Pakistani aquatic environment are agricultural runoff, household sewage release, and industrial effluents.
2. The levels of HM (viz., As, Hg, Cd, Pb, Cr, Ni, Fe, and Mn) found, often exceed MPL set by various domestic or international organizations, e.g., FAO/WHO, and ANMHRC/WAFDR.
3. We urge action to initiate a wider range of HM monitoring studies that will permit defining where HM exist in environmental media in Pakistan, and the degree of risk posed by such concentrations.

## 4 Summary

In this review, we evaluate and summarize the available data that addresses the levels of HM that exist in aquatic species, mainly fish, of Pakistan. Data on this topic were collected from the literature of the last two decades (1990–2012). Results revealed that the highest number (>50%) of studies addressing HM-contaminated fish have occurred in the Punjab province, followed by the Sindh and Khyber Pakhtunkhwa provinces. Our review disclosed that the HM concentrations in Pakistani fish species varied considerably with location. Generally, the level of HM residues detected in fish species had the following descending order: Fe > Zn > Pb > Cd > Hg > Ni > Cu > Ag > Cr > Mn > As. Fish samples collected from the Kabul River near the Nowshera district, Stretch of Ravi River, Indus River near Mainwali district, and Arabian Sea at Karachi revealed extremely high HM concentrations (range: 0.34–8,381.30 µg/g), compared to other fresh water bodies, such as the Llyold Barrage, Guddu Barrage, Jinnah Barrage, and Chashma Barrage (0.01–2.13 µg/g). As a reference point, we also reviewed selected data on HM fish residues that exist in countries that neighbor Pakistan. With the exception of fish collected in India, the majority of fish analyzed for HM residues in neighboring countries displayed lower residues than did fish from Pakistan.

We concluded from reviewing the available published data that the most probable sources for the HM contaminants found in Pakistani water and fish were release of domestic sewage, agricultural runoff, and industrial effluents. We strongly recommend that action be taken to better control the discharges of unregulated waste that enters the Pakistani aquatic environment, with the intent to mitigate any continuing future damage to the aquatic ecosystem. We also recommend intensifying research programs that address the toxicity of HM to the aquatic environment, so that a better understanding of metal effects on fish can be achieved that will lead to a sustainable ecological harmony in Pakistan.

**Acknowledgement** We would like to thank and highly acknowledge the efforts of Dr. David M. Whitacre, Summerfield, NC, USA, for his time to comment on the structure and English of the manuscript.

# References

Ahmad MS, Bibi S (2010) Uptake and bioaccumulation of water borne lead (Pb) in the fingerlings of a freshwater Cyprinid, *Catla catla*. J Anim Plant Sci 20(3):201–207

Ahmad T, Kahlown MA, Tahir A, Rashid H (2004) Arsenic an emerging issue, experiences from Pakistan. In: 30th WEDC international conference, Vientiane, Lao PDR

Alhashemi AH, Sekhavatjou MS, Kiabi BH, Karbassi AR (2012) Bioaccumulation of trace elements in water, sediment, and six fish species from a freshwater wetland, Iran. Microchem J 104:1–6

Amin MN, Begum A, Mondal MGK (2011) Trace element concentrations present in five species of freshwater fish of Bangladesh. Bangladesh J Sci Ind Res 46:27–32

Ansari TM, Saeed MA, Raza A, Naeem M, Salam A (2006) Effect of body size on metal concentration in wild Puntius chola. Pak J Environ Anal Chem 7(2):116–119

Arian MB, Kazi TG, Jamali MK, Jalbani N, Afridi HI, Shah A (2008) Total dissolved and bioavailable elements in water and sediment samples and their accumulation in *Oreochromis mossambicus* of polluted Manchar Lake. Chemosphere 70:1845–1856

Ashraf M, Tariq J, Jaffer M (1991) Content of trace metals in fish, sediments and water from three freshwater reservoirs on the Indus River Pakistan. Fish Res 12:355–364

Balkas IT, Tugrel S, Salhogln I (1982) Trace metal level in fish and Crustacea from the Northeastern Mediterranean waters. Mar Environ Res 6:281–289

Bianchi GW (1985) FAO species identification sheets for fishery purposes. Field guide to the commercial marine and brackish-water species of Pakistan. Prepared with support of PAK/77/033 and FAO (FIRM) Regular Programme. FAO, Rome

Cardwell RD, DeForest DK, Brix KV, Adams WJ (2013) Do Cd, Cu, Ni, Pb, and Zn biomagnify in aquatic ecosystems? Rev Environ Contam Toxicol 226:101–121

Chang S, Zdanowicz VS, Murchelano RA (1998) Associations between liver lesions in winter flounder (*Pleuronectes americanus*) and sediment chemical contaminants from north-east United States estuaries. ICES J Mar Sci 55:954–969

Chaudhry AS, Jabeen F (2011) Assessing metal, protein, and DNA profiles in *Labeo rohita* from the Indus River in Mianwali, Pakistan. Environ Monit Assess 174:665–679

Dallinger R, Prosi F, Segner H, Black H (1987) Contaminated food and uptake of heavy metals by rainbow trout (*Salmo gairdneri*): a field study. Oecologia 73:91–98

Demirak A, Yilmaz F, Tuna AL, Ozdemir N (2006) Heavy metals in water, sediment and tissues of *Leuciscus cephalus* from a stream in south western Turkey. Chemosphere 63:1451–1458

Dewani VK, Ansari IA, Khuhawar MY (2003) Profile of metal contents of sewage contaminated canal fishes. J Chem Soc Pak 25(1):11–15

Dix HM (1981) Environmental pollution. Wiley, New York

Duffus JH (1980) Environmental toxicology. Edward Arnold Publications Ltd., London

Ebrahimi M, Taherianfard M (2010) Concentration of four heavy metals (cadmium, lead, mercury, and arsenic) in organs of two cyprinid fish (Cyprinus carpio and Capoeta sp.) from the Kor River (Iran). Environ Monit Assess 168:575–585

FAO (1983) Compilation of legal limits for hazardous substances in fish and fishery products. FAO Fish Circ 464:5–100

Farooqi A, Masuda H, Firdous N (2007) Toxic fluoride and arsenic contaminated groundwater in the Lahore and Kasur districts, Punjab, Pakistan and possible contaminant sources. Environ Pollut 145:839–849

Fatoki OS, Mathabatha S (2001) An assessment of heavy metals pollution in the East London and Port Elizabeth Harbors. Water SA 27:233–236

Fu J, Hub X, Tao X, Yu H, Zhang X (2013) Risk and toxicity assessments of heavy metals in sediments and fishes from the Yangtze River and Taihu Lake, China. Chemosphere http://dx.doi.org/10.1016/j.chemosphere.2013.06.061

Gachal GS, Slater FM, Nisa Z, Qadri AH, Zuhra (2006) Ecological effect to the status of the Indus Dolphin. Pak J Biol Sci 9(11):2117–2121

Gummadavelli V, Piska RS, Noothi S, Manikonda PK (2013) Seasonal bioaccumulation of heavy metals in Cyprinus carpio of Edulabad water reservoir, Andhra Pradesh, India. Int J Life Sci Bt Pharm Res 2(3):127–143

Hajeb P, Jinap S, Ismail A, Mahyudin NA (2012) Mercury pollution in Malaysia. Rev Environ Contam Toxicol 220:45–66

Handy RD (1994) Intermittent exposure to aquatic pollutants assessment, toxicity and sublethal responses in fish and invertebrates. Comp Biochem Physiol 107c:184

Ittekkot V, Arian R (1986) Nature of particulate organic matter in the River Indus, Pakistan. Geochim Cosmochim Acta 50:1643–1653

Jabeen F, Chaudhry AS (2010) Environmental impacts of anthropogenic activities on the mineral uptake in *Oreochromis mossambicus* from Indus River in Pakistan. Environ Monit Assess 166:641–651

Jaffar M, Ashraf M, Rasool A (1988) Heavy metal contents in some selected local fresh water fish and relevant waters. Pak J Sci Ind Res 31:189–193

Jaffar M, Ashrif M, Tariq J (1995) Assessment of current trace metals pollution status of the Southeast Arabian Sea coast of Pakistan through fish analysis. J Chem Soc Pak 17(4):204–207

Jaleel T, Jaffar M, Ashraf M (1991) Levels of selected heavy metals in commercial fish from five fresh water lakes in Pakistan. Toxicol Environ Chem 33:133–140

Javed M, Usmani N (2011) Accumulation of heavy metals in fishes: a human health concern. Int J Environ Sci 2(2):659–670

Javid M (2003) Relationship among water, sediments and plankton for the uptake and accumulation of heavy metals in the River Ravi. Int J Plant Sci 2:326–331

Javid M (2005) Heavy metal contamination of freshwater fish and bed sediments in the River Ravi stretch and related tributaries. Pak J Biol Sci 8(10):1337–1341

Javid A, Javid M, Abdullah S, Ali Z (2007) Bio-accumulation of lead in the bodies of major carps (*Catla catla, Labeo rohita and Cirrhina mrigala*) during 96-h LC 50 exposures. Int J Biol 9(6):909–912

Khan T, Muhammad S, Khan B, Khan H (2011) Investigating the levels of selected heavy metals in surface water of Shah Alam River (a tributary of River Kabul, Khyber Pakhtunkhwa). J Himalaya Earth Sci 44(2):71–79

Khan B, Khan H, Muhammad S, Khan T (2012) Heavy metals concentration trends in three fish species from Shah Alam River, Khyber Pakhtunkhwa Province, Pakistan. J Nat Environ Sci 1:1–8

Khan MU, Malik RN, Muhammad S (2013) Human health risk from heavy metal via food crops consumption with wastewater irrigation practices in Pakistan. Chemosphere 93(10):2230–2238, http://dx.doi.org/10.1016/j.chemosphere.2013.07.067

Korai AL, Sahato GA, Kazi TG, Lashari KH (2008) Lead concentrations in fresh water, muscle, gill and liver of *Catla catla* (Hamilton) from Keenjhar lake. Pak J Anal Environ Chem 9(1):1119

Lyle JM (1984) Mercury concentrations in four carcharhinid and three hammerhead sharks from coastal waters of the Northern Territory. Aust J Mar Freshw Res 20:679–685

Malik RN, Nadeem M (2010) Spatial and temporal characterization of trace elements and nutrients in Rawal lake reservoir, Pakistan using multivariate analysis techniques. Environ Geochem Health 33(6):525–541

Malik RN, Moeckel C, Hughes D (2011) Polybrominated diphenyl ethers (PBDEs) in feather of colonial water birds species from Pakistan. Environ Pollut 159(10):3044–3050

Marcovecchio JE, Moreno VJ, Perez A (1986) Bio-magnification of total mercury in Bahia Blanca estuary shark. Mar Pollut Bull 17:276–278

Muhammad S, Shah MT, Khan S (2010) Arsenic health risk assessment in drinking water and source apportionment using multivariate statistical techniques in Kohistan region, northern Pakistan. Food Chem Toxicol 48:2855–2864

Muhammad S, Shah MT, Khan S (2011a) Health risk assessment of heavy metals and their source apportionment in drinking water of Kohistan region, northern Pakistan. Microchem J 98:334–343

Muhammad S, Shah MT, Khan S (2011b) Heavy metal concentrations in soil and wild plants growing around Pb-Zn sulfide terrain in Kohistan region, northern Pakistan. Microchem J 99:67–75

Muhammad S, Shah MT, Khan S, Saddique U, Gul N, Khan MU, Malik RN, Farooq M, Naz A (2013) Wild plant assessment for heavy metal phytoremediation potential along the mafic and ultramafic terrain in northern Pakistan. Biomed Res Int 2013:1–9, http://dx.doi.org/10.1155/2013/194765

Nawaz S, Nagra SA, Saleem Y, Priydarshi A (2010) Determination of heavy metals in fresh water fish species of the River Ravi, Pakistan compared to farmed fish varieties. Environ Monit Assess 167:461–471

Nussey G, Van Vuren JHJ, Du Preez HH (2000) Bioaccumulation of chromium, manganese, nickel and lead in the tissues of the moggel, *Labeo umbratus* (Cyprinidae), from Witbank Dam, Mpumalanga. Water SA 26(2):269–284

Paul MJ, Meyer JL (2001) Streams in the urban landscape. Annu Rev Ecol Syst 32:333–365

Plaskett D, Potter IC (1979) Heavy metal concentrations in the muscle tissue of 12 species of teleost from Cockburn Sound, Western Australia. Aust J Mar Freshw Res 30:607–616

Qadir A, Malik RN (2011) Heavy metals in eight edible fish species from two polluted tributaries (Aik and Palkhu) of the River Chenab, Pakistan. Biol Trace Elem Res 143:1524–1540

Qadir A, Malik RN, Husain SZ (2008) Spatio-temporal variations in water quality of Nullah Aik-tributary of the River Chenab, Pakistan. Environ Monit Assess 140:43–59

Qiu YW, Lin D, Liu JQ, Zeng EY (2011) Bioaccumulation of trace metals in farmed fish from South China and potential risk assessment. Ecotoxicol Environ Saf 74:284–293

Rahman MS, Molla AH, Saha N, Rahman A (2012) Study on heavy metals levels and its risk assessment in some edible fishes from Bangshi River, Savar, Dhaka, Bangladesh. Food Chem 134:1847–1854

Rashed MN (2001) Cadmium and lead levels in fish (*Tilapia nilotica*) tissues as biological indicator for lake water pollution. Environ Monit Assess 68(1):75–89

Rauf A, Javed M, Ubaidullah M (2009) Heavy metal levels in three major carps (*Catla catla*, *Labeo rohita* and *Cirrhina mrigala*) from the River Ravi Pakistan. Pak J Vet 29(1):24–26

Sarkar D, Datta R (2004) Arsenic fate and bioavailability in two soils contaminated with sodium arsenate pesticide: an indication study. Bull Environ Contam Toxicol 72:240–247

Shah AQ, Kazi TG, Arain MB, Baig JA, Afridi HI, Kandhro GA, Khan S, Jamalib M (2009) Hazardous impact of arsenic on tissues of same fish species collected from two ecosystems. J Hazard Mater 167:511–515

Shah AQ, Kazmi TG, Baig JA, Afridi HI, Khandhro GA, Khan S, Kolachi NF, Wadhwa SK (2010) Determination of total mercury in muscle tissues of marine fish species by ultrasonic assisted extraction followed by cold vapor atomic absorption spectrometry. Pak J Anal Environ Chem 11(2):12–17

Shah MT, Ara A, Muhammad S, Khan S, Tariq S (2012) Health risk assessment via surface water and sub-surface water consumption in the mafic and ultramafic terrain, Mohmand agency, northern Pakistan. J Geochem Explor 118:60–67

Shah MT, Ara J, Muhammad S, Khan S, Asad SA, Ali L (2013) Potential heavy metal accumulation of indigenous plant species along the mafic and ultramafic terrain in the Mohmand agency, Pakistan. Clean. doi:10.1002/clen.201200632

Shahid M, Pourrut B, Dumat C, Winterton P, Pinelli E (2011) Lead uptake, toxicity, and detoxification in plants. Rev Environ Contam Toxicol 213:113–136

Shahid M, Ferrand E, Schreck E, Dumat C (2013) Behavior and impact of zirconium in the soil-plant system: plant uptake and phytotoxicity. Rev Environ Contam Toxicol 221:107–127

Swaileh KM, Sansur R (2006) Monitoring urban heavy metal pollution using the house sparrow (*Passer domesticus*). J Environ Monit 8:209–213

Tabinda AB, Hussain M, Ahmed I, Yasar A (2010) Accumulation of toxic and essential trace metals in fish and prawns from Keti Bunder Thatta District, Sindh. Pak J Zool 42(5):631–638

Tariq J, Jaffar M, Ashraf M, Moazzam M (1991a) Heavy metals concentration in fish, shrimp, seaweed, sediment and water from the Arabian Sea, Pakistan. Mar Pollut Bull 26(11):644–647

Tariq J, Jaffar M, Moazzam M (1991b) Concentration correlations between major cations and heavy metals in fish from the Arabian Sea. Mar Pollut Bull 22(11):562–565

Tariq J, Jaffar M, Ashraf M (1992) Relationship between mercury concentration and length, weight and sex of two sprinid fish, *Catla catla and Chela chanius* from Rawal Lake Pakistan. Fish Res 14:335–341

Tariq J, Jaffar M, Ashraf M, Moazzam M (1993) Heavy metal concentrations in fish, shrimp, seaweed, sediment, and water from the Arabian Sea, Pakistan. Mar Pollut Bull 26(11):644–647

Tariq J, Jaffar M, Ashraf M (1994) Trace metal concentration, distribution and correlation in water, sediment and fish from the Ravi River, Pakistan. Fish Res 19:131–139

Tariq J, Ashraf M, Jaffar M, Afzal M (1995) Pollution status of Indus River Pakistan, through heavy metals and macronutrient content of fish sediments and water. Water Res 30(6):1837–1844

Tariq SS, Ashraf M, Jaffar M, Afzal M (1996) Pollution status of the Indus River, Pakistan, through heavy metals and macronutrient contents of fish, sediment and water. Water Res 30:1337–1344

Tariq J, Ashraf M, Jaffar M, Masud K (1998) Selected trace metals concentration in seven fish species from the Arabian Sea, Pakistan. J Chem Soc Pak 20(4):249–251

Vuren VJHJ, Preez HHD, Wepener V, Adendorfe A, Barnhroom IEJK (1999) Lethal and sub-lethal effects of metals on the physiology of fish. An experimental approach with monitoring support, WRC Report. Water Research Commission, Pretoria. 608/11/99

Wallace JRGT, Hoffman GL, Duce RA (1977) The influences of organic matter and atmospheric deposition on the particulate trace metal concentration of northwest Atlantic surface water. Mar Chem 5:143–170

WHO/FAO (1989) Evaluation of certain food additives and contaminants, FAO/WHO expert committee on food additives: 33rd Report Tech. Rep. Ser. 776. World Health Organization, Geneva

Widianarko B, VanGestel CAM, Verweij RA, Van-Straalen NM (2000) Associations between trace metals in sediment, water and guppy, *Poecilia reticulata* (Peters), from urban streams of Semarang, Indonesia. Ecotoxicol Environ Saf 46:101–107

Yousafzai AM, Khan AR, Shakoori AR (2008) Metal accumulation in the gills of an endangered South Asian fresh water fish as an indicator of aquatic pollution. Pak J Zool 40(5):331–339

Yousafzai AM, Chivers DP, Khan AR, Ahmad F, Siraj M (2010) Comparison of heavy metals burden in two freshwater fishes *Wallago attu* and *Labeo dyocheilus* with regards to their feeding habits in natural ecosystem. Pak J Zool 42(5):537–544

Yousafzai AM, Siraj M, Ahmad B, Chivers DP (2012) Bioaccumulation of heavy metals in common crap: implications for human health. Pak J Zool 44(2):489–494

Zhang G, Chakrabotry P, Li J, Sampathkumar P, Balasubramanian T, Kathiresan K, Takahashi S, Subramanian A, Tanabe S, Jones KC (2008) Passive atmospheric sampling of organochlorine pesticides, polychlorinated biphenyls, and polybrominated diphenyl ethers in urban, rural and wetland sites along the coastal length of India. Environ Sci Technol 42:818–822

# Index

INDEX FOR RECT VOLS. 221 to 230

17-[Beta]-estradiol (E2), predicted no-effect concentration (PNEC) in Chinese waters, **228**: 29 ff.
8-Hydroxyquinoline, acid mine drainage control, **226**: 17
9,10-Dihydro-9-oxa-10-phosphaphenanthrene-10-oxide, flame retardant, **222**: 42
9,10-Dihydro-9-oxa-10-phosphaphenanthrene-10-oxide, toxic & environmental properties, **222**: 43
9,10-Dihydro-9-oxa-10-phosphaphenanthrene-10-oxide, toxic & physical-chemical properties (table), **222**: 44
Abbreviations, defined (table), **227**: 57
Abiotic conditions, landfills & phthalates, **224**: 39 ff.
Abiotic degradation, **227**: 15
Abiotic degradation, fenpropathrin, **225**: 85
Abiotic degradation, methomyl, **222**: 98
Abiotic degradation, phthalate esters, **224**: 43
Absorption and transport, addictive substances, **227**: 66
Absorption, mercury, **229**: 5
Abuse by pregnant women, addictive substances (diag.), **227**: 56
Abused substance analysis value, for mothers, newborns and fetuses (diag.), **227**: 68
Abused substance biomonitoring, metabolites & biomarkers (table), **227**: 68–9
ACC (1-aminocyclopropane-1-carboxylic acid) deaminase, role in plant stress reduction, **223**: 44
ACC, plant root ethylene effect (diag.), **223**: 45

Accessory reproductive organs, bisphenol A effects, **228**: 72
Accumulation levels of copper, in land snails (table), **225**: 107
Accumulation processes, chemical elimination role, **227**: 111
Acetoanilide pesticides, leaching characteristics, **221**: 73
Acetoanilide pesticides, soil column leaching (table), **221**: 62–65
Acid & base hydrolysis, phthalates (diag.), **224**: 45
Acid mine drainage control, 8-hydroxyquinoline, **226**: 17
Acid mine drainage control, alkaline materials, **226**: 13
Acid mine drainage control, alkoxysilanes, **226**: 16
Acid mine drainage control, bactericides, **226**: 9
Acid mine drainage control, BDET, **226**: 17
Acid mine drainage control, catechol, **226**: 18
Acid mine drainage control, desulfurization, **226**: 19
Acid mine drainage control, electrochemical cover, **226**: 18
Acid mine drainage control, fatty acids, **226**: 17
Acid mine drainage control, humic acid, **226**: 15
Acid mine drainage control, inorganic coatings, **226**: 9
Acid mine drainage control, lipid use, **226**: 15
Acid mine drainage control, organic coatings, **226**: 15
Acid mine drainage control, oxalic acid, **226**: 17
Acid mine drainage control, passivation, **226**: 9
Acid mine drainage control, phosphate coating, **226**: 9

Acid mine drainage control, physical barriers, **226**: 4
Acid mine drainage control, polyethylene polyamine, **226**: 16
Acid mine drainage control, silica treatment, **226**: 11
Acid mine drainage locations, worldwide, **226**: 2
Acid mine drainage prevention, organic matter, **226**: 8
Acid mine drainage prevention, physical barriers, **226**: 5
Acid mine drainage, associated coal deposits (table), **226**: 3
Acid mine drainage, associated with sulfide deposits (table), **226**: 3
Acid mine drainage, control strategies, **226**: 4
Acid mine drainage, described, **226**: 2
Acid mine drainage, dry cover barriers, **226**: 6
Acid mine drainage, environmental problem, **226**: 2
Acid mine drainage, hazards, **226**: 3
Acid mine drainage, mineral associations (table), **226**: 3
Acid mine drainage, mitigation, **226**: 1 ff.
Acid mine drainage, prevention measures (diag.), **226**: 5
Acid mine drainage, prevention strategies, **226**: 5
Acid mine drainage, sulfide mineral oxidation, **226**: 2
Acid mine drainage, water cover, **226**: 6
Acidic effluent wastes, dry cover prevention (diag.), **226**: 7
Acidic mine effluent prevention, polymer barriers, **226**: 7
Acute toxicity to aquatic species, diazinon, **223**: 130
Acute toxicity to freshwater aquatic species, diazinon (table), **223**: 131
Addicted mothers, children effects, **227**: 58
Addiction implication, pregnant women, **227**: 58
Addiction monitoring, options & approaches, **227**: 56
Addiction severity index, described, **227**: 59
Addictive substance abuse, problem magnitude, **227**: 58
Addictive substance analysis, in biological media, **227**: 67
Addictive substance analysis, target biological materials (table), **227**: 70–3
Addictive substance biomonitoring, placenta role, **227**: 64

Addictive substance effects, during pregnancy & infancy, **227**: 55 ff.
Addictive substance effects, on mothers, fetuses & newborns (table), **227**: 60–3
Addictive substance exposures, prenatal effects, **227**: 59
Addictive substance use, by pregnant women, **227**: 58
Addictive substances abuse, by pregnant women (diag.), **227**: 56
Addictive substances analysis, in biological materials, **227**: 55 ff.
Addictive substances, abuse & prevalence, **227**: 55
Addictive substances, defined, **227**: 58
Addictive substances, *in utero* biomonitoring, **227**: 63
Addictive substances, maternal absorption & transport, **227**: 66
Addictive substances, metabolism, **227**: 64
Adult asthma causes, human exposure pathways (diag.), **226**: 35
Adult asthma influence, air pollution, **226**: 46
Adult asthma prevalence, rural vs. urban areas, **226**: 36
Adult asthma, changing rates, **226**: 34
Adult asthma, combustion products, **226**: 37
Adult asthma, environmental tobacco smoke, **226**: 37
Adverse effects, phthalate esters, **224**: 42
Aerobic soil metabolism, trace elements, **225**: 19
Aflatoxins, drinking water contaminant, **228**: 122
Agbogbloshie (Ghana) scrap yard, e-waste water pollution, **229**: 22
Agbogbloshie (Ghana), e-waste recycling (illus.), **229**: 23
Agricultural films, PBM (polymer-based materials) waste, **227**: 9
Agricultural vs. urban use in California, diazinon (table), **223**: 113
Air behavior, diazinon, **223**: 126
Air behavior, fenpropathrin, **225**: 82
Air behavior, methomyl, **222**: 98
Air contaminants, categories, **223**: 3
Air contaminants, gases, **223**: 5
Air contaminants, health effects, **223**: 5
Air contaminants, heavy metals, **223**: 7
Air contaminants, persistent organic pollutants (POPs), **223**: 7
Air contaminants, Santiago, Chile, **223**: 1ff.
Air contaminants, statistical distributions, **223**: 8

Index 135

Air contaminants, suspended solid particles, **223**: 7
Air contamination studies, based on Birnbaum-Saunder's models (table), **223**: 13
Air contamination studies, Santiago, Chile (table), **223**: 13
Air contamination, Santiago, Chile, **223**: 12
Air contamination, worldwide (table), **223**: 9–11
Air degradation pathway, diazinon (diag.), **223**: 127
Air exposure, copper, **225**: 100
Air monitoring locations, Santiago, Chile (illus.), **223**: 15
Air monitoring, diazinon in California, **223**: 128
Air pollutant characteristics, Santiago, Chile, **223**: 17
Air pollutant concerns, the Americas, **223**: 2
Air pollutant effects, Santiago, Chile, **223**: 15
Air pollutant hazards, Chilean guidelines (table), **223**: 16
Air pollutant monitoring, Santiago, Chile, **223**: 14
Air pollutant, Hg from stack emissions, **230**: 2
Air pollutants, maximum permitted levels (table), **223**: 16
Air pollutants, particulate matter, **223**: 3
Air pollution data, statistical analysis, **223**: 17
Air pollution data, statistical models, **223**: 3
Air pollution study, Santiago, Chile, **223**: 16
Air pollution, health effects, **223**: 2
Air pollution, influence on adult asthma, **226**: 46
Air pollution, people effects, **226**: 34
Air pollution, Santiago, Chile, **223**: 2
Alflatoxin biosynthesis-clustered genes, fungi (diag.), **228**:128
Alien species spread, PBM debris role, **227**: 27
Alkaline materials, acid mine drainage control, **226**: 13
Alkoxysilanes, acid mine drainage control, **226**: 16
Allergic disease incidence, from biological contaminant exposure, **226**: 43
Aluminum diethylphosphinate, flame retardant, **222**: 45
Aluminum diethylphosphinate, physical-chemical & environmental character, **222**: 45
Aluminum diethylphosphinate, toxic, physical-chemical & environmental properties (table), **222**: 46

Aluminum diethylphosphinate, toxicity, **222**: 48
Aluminum trihydroxide, chemical & environmental character (table), **222**: 11
Aluminum trihydroxide, environmental behavior, **222**: 9
Aluminum trihydroxide, inorganic flame retardant, **222**: 9
Aluminum trihydroxide, toxicity, **222**: 10
Alzheimer's disease link, mercury, **229**: 10
Alzheimer's disease, main features, **229**: 11
American kestrel, methyl mercury toxicity, **223**: 64
American kestrels, DE-71 (PBDE isomer) NOAEL & LOAEL (table), **229**: 125
Ammonium polyphosphate, chemical structure (illus.), **222**: 14
Ammonium polyphosphate, flame retardant, **222**: 13
Ammonium polyphosphate, physical-chemical properties (table), **222**: 15
Ammonium polyphosphate, toxicity & environmental character, **222**: 14–16
Amphibian toxicity, of 17-[beta]-estradiol (E2), **228**: 33
Amyotrophic lateral sclerosis, description, **229**: 11
Amyotrophic lateral sclerosis, mercury link, **229**: 6, 10
Anaerobic soil metabolism, trace elements, **225**: 19
Analysis and detection, mycotoxins, **228**:113
Analysis method, ptaquiloside, **224**: 76
Analysis of complex matrices, general scheme (diag.), **227**: 69
Analytical scheme, for biological sample analysis (diag.), **227**: 69
Androgen/estrogen synthesis & action effects, bisphenol A (BPA)(table), **228**: 66
Animal effects of metals, genetic diversity, **227**: 92
Animal energetics implication, bioamplification, **227**: 114
Animal energetics, lipid reserve management, **227**: 114
Animal genetic effects, metal pollution (table), **227**: 86–91
Animal health effects, boron, **225**: 59
Animal models, Bracken fern toxicity, **224**: 67
Animal populations, assessing genetic variability, **227**: 95
Animal populations, genetic diversity effects, **227**: 79 ff.
Animal populations, genetic structure effects, **227**: 79 ff.

Animal test results, Bracken fern intake (table), **224**: 68
Animal toxicity, boron, **225**: 60
Animal toxicity, Bracken fern intake, **224**: 67
Animals, boron intake & effects, **225**: 59
Animals, role of boron, **225**: 58
Anthropogenic emissions, zirconium, **221**: 109
Anthropogenic releases, mercury, **226**: 66
Anthropogenic sources, cadmium, **229**: 53
Anthropogenically-produced compounds, nanoparticles (table), **230**: 84
Antibacterial properties, silver compounds, **223**: 89
Aphagia effects, on bioamplification, **227**: 124
Apoptosis, ROS (reactive oxygen species) mode of action for nanosilver, **223**: 87
Applications described, fabricated nanoparticles, **230**: 85
Applications, fabricated nanoparticles (table), **230**: 86
Aquatic avian species in China, PBDE egg TRVs (diag.), **229**: 129
Aquatic biota, mercury residues, **226**: 77
Aquatic birds in China, PCBs (polychlorinated biphenyls), **230**: 59 ff.
Aquatic birds, PBDE TRV derivation, **229**: 121
Aquatic birds, PBDE TRV, **229**: 125
Aquatic Chinese species, setting WQC (water quality criteria), **230**: 38
Aquatic ecosystems, do heavy metals biomagnify?, **226**: 101 ff.
Aquatic environment degradation, PBMs, **227**: 22
Aquatic environmental effects, e-waste disposal, **229**: 19 ff.
Aquatic environments, diazinon persistence, **223**: 125
Aquatic life criteria, pesticides in sediments, **224**: 97 ff.
Aquatic life effects, heavy metals, **229**: 26
Aquatic life sediment quality criteria, protection defined, **224**: 112
Aquatic life, organic pollutant effects, **229**: 28
Aquatic life-benchmarks, diazinon, **223**: 133
Aquatic mammals, PBDE TRV derivation, **229**: 115
Aquatic mammals, TRV for PBDEs, **229**: 119
Aquatic organism toxicity, nanosilver, **223**: 83
Aquatic organism toxicity, of E2, **228**: 31
Aquatic organism toxicity, PBDEs, **229**: 28
Aquatic organisms, e-waste contaminant effects, **229**: 25
Aquatic organisms, lead & cadmium toxicity, **229**: 27
Aquatic organisms, methomyl toxicity (table), **222**: 104
Aquatic photolysis, diazinon, **223**: 126
Aquatic residues, mercury, **226**: 68
Aquatic sediments, characterized, **224**: 98
Aquatic species acute toxicity, diazinon (table), **223**: 131
Aquatic species chronic toxicity, diazinon (table), **223**: 132
Aquatic species ecotoxicity, fenpropathrin, **225**: 89
Aquatic species in China, PBDE TRVs (diag.), **229**: 129
Aquatic species toxicity, diazinon, **223**: 130
Aquatic species toxicity, E2 (table), **228**: 34–5
Aquatic species toxicity, nanosilver, **223**: 93
Aquatic species, clomazone toxicity, **229**: 44
Aquatic species, E2 NOEC values (table), **228**: 38
Aquatic species, methomyl toxicity, **222**: 103
Aquatic species, risk assessments, **230**: 49
Aquatic toxicity, diazinon, **223**: 129
Aquatic toxicity, pesticides, **224**: 99
Aquatic-life benchmark values, diazinon (table), **223**: 133
Aqueous environments, field data on trophic transfer, **226**: 106
Arsenate metabolism, arsenate reductases, **224**: 15
Arsenate reductases, arsenate metabolism, **224**: 15
Arsenate, description, **224**: 2
Arsenic & chromium metabolism, rhizospheric soil effect (diag.), **225**: 29
Arsenic acid derivatives, arsenic metabolism, **224**: 9
Arsenic behavior, soils, **225**: 4
Arsenic contamination, bioremediation methods, **224**: 19–25
Arsenic cycle, global (diag.), **224**: 5
Arsenic cycle, microbial interactions, **224**: 1 ff.
Arsenic cycle, microbial transformations, **224**: 6
Arsenic detoxification, eukaryotic cells (diag.), **224**: 11
Arsenic forms, posttransductional regulation, **224**: 17
Arsenic in soil, uses (table), **225**: 5
Arsenic metabolism effect, soil amendment (diag.), **225**: 35
Arsenic metabolism, arsenic permeases, **224**: 16
Arsenic metabolism, influencing factors, **225**: 23
Arsenic metabolism, methylarsines, **224**: 9
Arsenic metabolism, soil microbes, **225**: 19
Arsenic permeases, arsenic metabolism **224**: 16

Index

Arsenic pollution, sources, **224**: 5
Arsenic trioxide, "Gasio-gas" poisoning, **225**: 3
Arsenic, ars operon & transcriptional regulation, **224**: 14
Arsenic, biosensors, **224**: 25
Arsenic, characteristics (table), **224**: 2
Arsenic, description & uses, **224**: 2
Arsenic, detoxification mechanism, **224**: 12
Arsenic, environmental distribution, **224**: 4
Arsenic, environmental fate, **224**: 4
Arsenic, environmental levels (table), **224**: 3
Arsenic, enzyme reduction, **224**: 7
Arsenic, measuring bioavailability, **224**: 25
Arsenic, methylation and demethylation, **224**: 8
Arsenic, microbial detoxification & transport (diag.), **224**: 9
Arsenic, microbial resistance, **224**: 6
Arsenic, microbial sensing, **224**: 13
Arsenic, microbial uptake & extrusion, **224**: 10
Arsenic, mobilization and immobilization, **224**: 10
Arsenic, soil sources & behavior, **225**: 4
Arsenic, the ars operon & proteins, **224**: 13
Arsenic, toxicity, **224**: 2
Arsenic-resistant microbes, examples (table), **224**: 6
Arsenic-resistant microbes, examples, **224**: 20
Arsenic-toxicity mitigation, organismal strategies, **224**: 20
Arsenite reduction, pathways (table), **224**: 8
Arsenite, oxidation processes, **224**: 7
Arylalkanoate pesticides, leaching characteristics, **221**: 73
Arylalkanoate pesticides, soil column leaching (table), **221**: 62–65
Asthma & smoking behavior, urban vs. rural differences, **226**: 37
Asthma cause, coal & biomass fuel, **226**: 39
Asthma cause, nitrogen dioxide exposure, **226**: 39
Asthma cause, particulate matter exposure, **226**: 39
Asthma effects, urban vs. rural residents (table), **226**: 53–55
Asthma in adults, atopy in rural vs. urban residents, **226**: 45
Asthma incidence, socioeconomic effects, **226**: 49
Asthma influence, climatic factors, **226**: 48
Asthma influence, geographic factors, **226**: 48
Asthma morbidity factors, biological contaminants, **226**: 40
Asthma morbidity, rural living benefit, **226**: 44

Asthma prevalence, healthcare access effects, **226**: 49
Asthma prevalence, urban vs. rural differences (table), **226**: 52
Asthma, described, **226**: 34
Atmospheric mercury, Mexico, **226**: 65 ff.
Atmospheric pollutants, Gaussian statistical modeling, **223**: 8
Atmospheric pollutants, worldwide (table), **223**: 9–11
Atmospheric release, mercury levels, **226**: 67
Atopy & asthma in adults, urban vs. rural effects, **226**: 45
Australian drinking water guidelines, chemical contaminants (table), **222**: 122
Australian guidelines, drinking water quality, **222**: 121
Australian trends, potable water disinfection, **222**: 127
Autocorrelation analysis, Santiago, Chile, air pollution data, **223**: 18
Autoimmune effects, mercury, **229**: 6
Avian species in China, compared (table), **223**: 56
Avian species in China, methyl mercury levels (diag.), **223**: 73
Avian species in China, TRV (toxicity reference value) & TRC (tissue residue criteria) derivation, **223**: 66
Avian species toxicity thresholds, methyl mercury, **223**: 65
Avian species toxicity, methyl mercury, **223**: 58
Avian species toxicity, selecting data for protection, **223**: 56
Avian species, body masses & food ingestion rates (table), **230**: 62
Avian sub-chronic and chronic toxicity, methyl mercury (table), **223**: 59–61
Avian wildlife protection in China, methyl mercury, **223**: 53 ff.
Avian wildlife TRV uncertainty factors, methyl mercury (table), **223**: 66
Avian wildlife TRVs & TRCs, methyl mercury (table), **223**: 67
Avian wildlife, Chinese species to protect, **223**: 55
Bacteria, boron effects, **225**: 62
Bacteria, plant growth effect, **223**: 40
Bacterial cell-wall interaction, silver nanoparticles, **223**: 90
Bacterial inhibition, acid mine drainage control, **226**: 8
Bacterial waterborne pathogens, in drinking water, **222**: 130

Bactericides, acid mine drainage control, **226**: 9
Bald eagle, PCB toxicity, **230**: 69
Bangladesh fish residues, heavy metals, **230**: 126
BDET (3-benzenediamidoethanthiol), acid mine drainage control, **226**: 17
Bioaccumulation effects, soil pH, **225**: 24
Bioaccumulation effects, trace elements, **225**: 36
Bioaccumulation mechanism, bioamplification, **227**: 107 ff., 108
Bioaccumulation mechanism, bioamplification, **227**: 109
Bioaccumulation of copper, by land snails, **225**: 111
Bioaccumulation of copper, land snails (table), **225**: 101
Bioaccumulation of HFFRs (halogen free flame retardants), classification scheme (table), **222**: 9
Bioaccumulation process, described, **226**: 102, **227**: 107
Bioaccumulation simulations, model parameters (table), **227**: 138–9, 143–5
Bioaccumulation, 9,10-Dihydro-9-oxa-10-phosphaphenanthrene-10-oxide, **222**: 43
Bioaccumulation, aluminum diethylphosphinate (table), **222**: 46
Bioaccumulation, ammonium polyphosphate (table), **222**: 15
Bioaccumulation, bisphenol-A bis(diphenylphosphate) (table), **222**: 40
Bioaccumulation, bisphenol-A bis(diphenylphosphate), **222**: 38
Bioaccumulation, diazinon, **223**: 132
Bioaccumulation, HFFRs (table), **222**: 59
Bioaccumulation, HFFRs, **222**: 1 ff.
Bioaccumulation, HFFRs, **222**: 58
Bioaccumulation, HFFRs, **222**: 6
Bioaccumulation, implication for deriving sediment quality criteria, **224**: 158
Bioaccumulation, melamine polyphosphate (table), **222**: 49
Bioaccumulation, melamine polyphosphate, **222**: 51
Bioaccumulation, pentaerythritol (table), **222**: 54
Bioaccumulation, pentaerythritol, **222**: 52
Bioaccumulation, resorcinol bis(diphenylphosphate) (table), **222**: 32
Bioaccumulation, resorcinol bis(diphenylphosphate), **222**: 36
Bioaccumulation, scientific value, **227**: 108
Bioaccumulation, trace elements, **225**: 7

Bioaccumulation, triphenylphosphate (table), **222**: 27
Bioaccumulation, triphenylphosphate, **222**: 30
Bioaccumulation, zinc hydroxystannate, **222**: 20
Bioaccumulation, zinc stannate, **222**: 22
Bioamplification effect, aphagia, **227**: 124
Bioamplification factor, described, **227**: 112
Bioamplification factors, herring gull modeling (table), **227**: 131
Bioamplification factors, yellow perch modeling (table), **227**: 129
Bioamplification modeling, herring gulls (diags.), **227**: 130
Bioamplification modeling, yellow perch (diags.), **227**: 127
Bioamplification of POPs (persistent organic pollutants), modeling approaches, **227**: 124
Bioamplification simulation, herring gull model, **227**: 141
Bioamplification simulation, yellow perch model, **227**: 137
Bioamplification, and lipid loss, **227**: 113
Bioamplification, animal energetics implication, **227**: 114
Bioamplification, bioaccumulation mechanism, **227**: 107 ff.
Bioamplification, case studies (table), **227**: 116–7
Bioamplification, case studies, **227**: 114
Bioamplification, data gaps identified, **227**: 135
Bioamplification, during metamorphosis & reproduction, **227**: 119
Bioamplification, during migration, **227**: 122
Bioamplification, during overwintering & hibernation, **227**: 121
Bioamplification, embryo & juvenile development, **227**: 115
Bioamplification, implication, **227**: 133
Bioamplification, independent bioaccumulation mechanism, **227**: 109
Bioamplification, measurement, **227**: 112
Bioamplification, summary conclusions, **227**: 135
Bioamplification, term defined & described, **227**: 108
Bioamplification, vs. other accumulation processes, **227**: 110
Bioamplification, weight-loss link, **227**: 111
Bioavailability considerations, in deriving sediment quality criteria, **224**: 135
Bioavailability in soil, metals, **223**: 37
Bioavailability in soil, trace elements, **225**: 1ff.
Bioavailability in soils, zirconium, **221**: 110

Index

Bioavailability of arsenic, measurement, **224**: 25
Bioavailability of metals, hyperaccumulator plant effect, **223**: 39
Biochar ageing effect, soil microbes, **228**: 93
Biochar effect, biomass source, **228**: 89
Biochar effect, soil pH, **228**: 90
Biochar effects, factors affecting performance (diag.), **228**: 90
Biochar elemental composition, biomass source effects (diag.), **228**: 88
Biochar elemental composition, processing effects (table), **228**: 92
Biochar performance effects, soil organic matter & minerals, **228**: 91
Biochar performance, pyrolysis temperature effects, **228**: 87
Biochar remediation performance, temperature effects, **228**: 93
Biochar, description & characteristics, **228**: 84
Biochar, for remediating contaminated soils, **228**: 83 ff., 85
Biochar, future research needs, **228**: 94
Biochar, organics & metal adsorption, **228**: 86–7
Biochar, pesticide & metal adsorption, **228**: 86
Biochar, properties and performance, **228**: 86
Bioconcentration, defined, **227**: 109
Bioconcentration, described, **226**: 102
Biodegradation effects, pesticide mobility, **221**: 23
Biodegradation rate, method effects, **221**: 24
Biodegradation, PBMs (polymer-based materials), **227**: 21
Biodegradation, phthalates, **224**: 47
Biodiversity effects, marine debris, **228**: 19
Biofilm formation, in drinking water, **222**: 144
Biogeochemical behavior effects, zirconium speciation, **221**: 113
Biogeochemical cycle, mercury, **226**: 66
Bioindicator organisms, described, **225**: 96
Biokinetic data for trophic transfer, accumulator marine species (diag.), **226**: 112
Biokinetic heavy metal data, freshwater trophic transfer factors (diag.), **226**: 110
Biokinetic marine data, heavy metal trophic transfer (diag.), **226**: 111
Biokinetic models, for estimating trophic transfer factors, **226**: 104
Biological attributes, land snails, **225**: 96
Biological contaminant exposure, allergic disease incidence, **226**: 43
Biological contaminants, adult asthma cause, **226**: 40

Biological effects, nanosilver, **223**: 81 ff.
Biological effects, of boron, **225**: 57 ff.
Biological material analysis, for addictive substances, **227**: 55 ff.
Biological materials selected, for addictive substance analysis (table), **227**: 70–3
Biological media analysis, for addictive substances, **227**: 67
Biological organization level to protect, sediment quality criteria, **224**: 112
Biological pollutants, drinking water, **228**:122
Biological sample analysis, analytical scheme (diag.), **227**: 69
Biological specimen analysis, of maternal, fetus & newborn samples (diag.), **227**: 68
Biomagnification of heavy metals, does it occur?, **226**: 101 ff.
Biomagnification, described, **226**: 102,**227**: 110
Biomarker in land snails, histopathology, **225**: 114
Biomarker in land snails, lysosomal stability, **225**: 117
Biomarker in land snails, membrane integrity, **225**: 117
Biomarker in land snails, ultrastructure, **225**: 114
Biomarker role, genetic damage monitoring, **227**: 84
Biomarkers in land snail, metallothionein induction, **225**: 118
Biomarkers in land snails, copper pollution, **225**: 113
Biomarkers in land snails, dependencies (table), **225**: 115
Biomarkers, abused substance biomonitoring (table), **227**: 68–9
Biomarkers, to detect genetic diversity effects, **227**: 85
Biomass fuel & coal, adult asthma cause, **226**: 39
Biomass source, biochar effect, **228**: 89
Biomass sources, biochar elemental composition (diag.), **228**: 88
Biomethylation, selenium, **225**: 20
Biomethylation, trace elements in soil, **225**: 17
Biomonitoring *in utero*, addictive substances, **227**: 63
Biomonitoring of abused substances, metabolites & biomarkers (table), **227**: 68–9
Biomonitoring of addictive substances, placenta role, **227**: 64
Bioremediation methods, arsenic contamination, **224**: 19–25

Bioremediation, definition, **223**: 35, **224**: 20
Bioremediation, of metalloid contamination, **225**: 38
Biosensors, arsenic, **224**: 25
Biosorption of trace elements, in aquatic environment (table), **225**: 9
Biosorption processes, trace elements, **225**: 11
Biosorption, trace elements (table), **225**: 9
Biota, copper exposure routes, **225**: 97
Biotic degradation, fenpropathrin, **225**: 84
Biotic degradation, methomyl, **222**: 101
Biotic degradation, PBMs, **227**: 21
Biotic processes, chlomazone, **229**: 40
Bird ecotoxicity, fenpropathrin, **225**: 90
Bird residues in Mexico, mercury, **226**: 84
Bird residues, of PCBs (table), **230**: 76
Bird tissue residues, mercury (table), **226**: 85
Bird toxicity studies, PCBs, **230**: 64
Bird toxicity, methomyl, **222**: 104
Bird toxicity, of DE-71 (PBDE isomer), **229**: 121
Bird toxicity, PCB mixtures and isomers (table), **230**: 65–67
Birds, PBDE TRVs (table), **229**: 127
Birds, PCB risk assessment, **230**: 76
Birnbaum-Saunder's models, air contaminant distribution, **223**: 12
Bisphenol A derivatives, male reproductive effects, **228**: 72
Bisphenol A effects, on hypothalamic-pituitary-testicular axis, **228**: 62
Bisphenol A effects, on sperm function, **228**: 70
Bisphenol A effects, testicular & epididymal antioxidant system, **228**: 70
Bisphenol A *in utero* exposure, male reproductive effects, **228**: 6–20
Bisphenol A, accessory reproductive organs, **228**: 72
Bisphenol A, described, **228**: 58
Bisphenol A, exposure sources & routes, **228**: 58
Bisphenol A, human residues, **228**: 59
Bisphenol A, male reproductive effects, **228**: 57 ff.
Bisphenol A, occupational exposure, **228**: 60
Bisphenol A, spermatogenesis effects, **228**: 62, 68
Bisphenol-A bis(diphenylphosphate), flame retardant, **222**: 37
Bisphenol-A bis(diphenylphosphate), physical-chemical & environmental properties, **222**: 38

Bisphenol-A bis(diphenylphosphate), toxic, environmental & physical-chemical properties (table), **222**: 39–41
Bisphenol-A bis(diphenylphosphate, structure (illus.), **222**: 38
Body masses, avian species (table), **230**: 62
Bootstrapping, constructing species sensitivity distributions, **230**: 41
Boron & borax, rat subchronic toxicity, **225**: 68
Boron deficiency effects, plants, **225**: 62
Boron effects, bacteria & fungi, **225**: 62
Boron effects, on enzymes & minerals, **225**: 63
Boron in cattle, herpes virus resistance, **225**: 65
Boron intake & effects, animals, **225**: 59
Boron intake & sources, humans, **225**: 61
Boron intake effects, humans, **225**: 61
Boron regulatory role, minerals, **225**: 65
Boron toxicity, human fatality, **225**: 67
Boron toxicity, humans, **225**: 66
Boron toxicity, in animals, **225**: 60
Boron vs. borax, toxicity, **225**: 66
Boron, acute toxicity in rats, **225**: 67
Boron, biological effects, **225**: 57 ff.
Boron, chicken toxicity, **225**: 60, 67
Boron, clinical effects in animals, **225**: 59
Boron, description & characteristics, **225**: 57
Boron, natural chemical forms, **225**: 58
Boron, natural sources, **225**: 58
Boron, plant toxicity, **225**: 69
Boron, role in animals, **225**: 58
Boron, role in living organisms, **225**: 58
Boron, role in plants, **225**: 61
Boron, toxicity symptoms in humans, **225**: 67
Boron, toxicity, **225**: 66
Bovine papillomavirus amplification, polymerase chain reaction, **224**: 83
Bovine papillomavirus diagnosis, polymerase chain reaction role, **224**: 85
Bovine papillomavirus role, tumor induction & progression, **224**: 64
Bovine papillomavirus, Bracken fern interaction, **224**: 65
Bovine papillomavirus, described, **224**: 64
Bovine papillomavirus, tumor progression & Bracken fern, **224**: 64
Bovine tumor classification, pathology, **224**: 66
BPA (bisphenol A), environmental concentrations, **227**: 33
BPA (bisphenol A), male reproductive effects (table), **228**: 63
BPA action sites, male reproductive function (diag.), **228**: 71
BPA, androgen/estrogen synthesis & action effects (table), **228**: 66

Index 141

BPA, testicular effects, **228**: 65
BPA, tolerable daily intake values, **228**: 73
Bracken & other ferns, cause of EBH (enzootic bovine hematuria) in cattle, **224**: 57
Bracken fern carcinogen, ptaquiloside, **224**: 57
Bracken fern effects, animals (table), **224**: 68
Bracken fern interaction, bovine papillomavirus, **224**: 65
Bracken fern plant, photos (illus.), **224**: 58
Bracken fern toxicity, animal models, **224**: 67
Bracken fern toxicity, animals, **224**: 67
Bracken fern, gene mutations, **224**: 79
Bracken fern, oxidative stress role, **224**: 82
Brazilian coastal waters, DDT contamination, **228**: 8
Brominated flame retardants, applications & alternatives thereto (table), **222**: 4
Brominated flame retardants, described, **222**: 2
Brominated flame retardants, uses curtailed, **222**: 3
Bulk PBMs, environmental release, **227**: 7
Cadmium effects, chlorophyll & carotenoids, **229** 60, 61
Cadmium effects, on fatty acid composition, **229**: 63
Cadmium effects, on free amino acids & proline, **229**: 62
Cadmium effects, on genetic material, **229**: 63
Cadmium effects, on key organic osmotica, **229**: 65
Cadmium effects, on plant growth (table), **229**: 56
Cadmium effects, on protein levels, **229**: 61
Cadmium effects, photosynthesis, **229**: 69
Cadmium effects, plant growth regulators, **229**: 67
Cadmium effects, plant metal uptake, **229**: 64
Cadmium toxicity, to aquatic organisms, **229**: 27
Cadmium, described, **229**: 52
Cadmium, environmental sources, **229**: 52
Cadmium, natural vs. anthropogenic sources, **229**: 52
Cadmium, oxidative defense system effects, **229**: 66
Cadmium, plant accumulation, **229**: 54
Cadmium, plant effects, **229**: 59
Cadmium, plant morphology effects, **229**: 55
Cadmium, plant pigment effects, **229**: 59
Cadmium, plant water relations effect, **229**: 70
Cadmium, protein and protease effects, **229**: 61
Cadmium, soil and plant behavior, **229**: 54
Cadmium-enriched environments, plant growth effects, **229**: 51 ff.

Cadmium-enriched environments, plant metabolic effects, **229**: 51 ff.
California agricultural counties, diazinon water monitoring data (table), **223**: 124
California annual insecticide use, diazinon (illus.), **223**: 111
California counties, highest diazinon use, **223**: 111
California counties, major diazinon use (table), **223**: 112
California crop use, fenpropathrin (diag.), **225**: 80
California pesticides, types used, **224**: 99
California top crop uses, diazinon (diag.), **223**: 113
California urban watersheds, diazinon residues (table), **223**: 124
California use by application method, diazinon (diag.), **223**: 114
California use profile, diazinon, **223**: 107, 110
California use volume (2000–2010), fenpropathrin (diag.), **225**: 79
California, county map (illus.), **223**: 110
California, fenpropathrin weight used (diag.), **225**: 80
Canadian drinking water guidelines, microbial & chemical contaminants (tables), **222**: 121
Canadian guidelines, drinking water quality, **222**: 120
Canadian trends, potable water disinfection, **222**: 126
Cancer diagnostic aid, ultrasonography, **224**: 82
Cancer-promoting metabolites, fumonisins, **228**: 106
Carbamate pesticides, soil column leaching results (table), **221**: 54–57
Carbamate pesticides, soil leaching results, **221**: 53
Carcinogenic mechanism, ptaquiloside (diag.), **224**: 63
Carcinogenic mechanism, ptaquiloside, **224**: 62
Cardiac laboratories, radiation exposure routes (illus.), **222**: 87
Cardiac medical procedures, described, **222**: 74
Cardiology staff effects, radiation exposure, **222**: 85
Cardiology staff exposure, radiation, **222**: 73 ff.
Cardiology staff radiation dose, eyes & forehead (diag.), **222**: 80
Cardiology staff radiation exposure, cataract formation, **222**: 84
Cardiology staff radiation exposure, eye & thyroid gland, **222**: 82

Cardiology staff radiation exposure, literature searched, **222**: 75
Cardiology staff, health effects from radiation, **222**: 73 ff.
Cardiology staff, radiation dose by procedure (diag.), **222**: 79
Cardiology staff, radiation dose to hands (diag.), **222**: 82
Cardiology staff, radiation dose to thyroid/neck region (illus.), **222**: 81
Cardiology staff, radiation doses, **222**: 78
Cardiology staff, radiation hand and wrist exposure, **222**: 80
Cardiology staff, radiation safety practices, **222**: 85
Carotenoid effects, cadmium, **229**: 60
Case studies described, bioamplification, **227**: 114
Cataract formation, irradiated cardiology staff, **222**: 84
Catechol, acid mine drainage control, **226**: 18
Catheterization laboratories, radiation exposure routes (illus.), **222**: 87
Catheterization laboratory, growth & trends, **222**: 76
Catheterization safety practices, radiation, **222**: 85
Cd biomagnification, in aquatic ecosystems, **226**: 101 ff.
Cd residues, in Guiana dolphin kidney & liver (table), **228**: 14
Cereal levels, *Fusarium* toxins (table), **228**: 104
Cerium nanoparticles, described, **230**: 99
Cetacean liver, mercury contamination (diag.), **228**: 12
Cetacean threat, endocrine disruptors, **228**: 2
Cetaceans, DDT & PCB residues (diag.), **228**: 8
Cetaceans, pollutant accumulation, **228**: 2
Chemical additives to plastics, toxicity, **227**: 34
Chemical behavior, elemental mercury, **229**: 2
Chemical behavior, phthalates, **224**: 39 ff.
Chemical compounds, placental transport (diag.), **227**: 67
Chemical contaminants in drinking water, EU Directive (table), **222**: 123
Chemical contaminants, Australian drinking water guidelines (table), **222**: 122
Chemical contaminants, Canadian drinking water guidelines (table), **222**: 121
Chemical contamination, drinking water guidelines (table), **222**: 116
Chemical contamination, in cetaceans, **228**: 2
Chemical damage to DNA, implications, **227**: 83

Chemical degradation, microbe involvement, **221**: 25
Chemical elimination, in accumulation processes, **227**: 111
Chemical forms & sources, boron, **225**: 58
Chemical forms, mercury, **229**: 2
Chemical mixtures, effect on deriving sediment quality criteria, **224**: 155
Chemical standards, for drinking water (table), **222**: 117
Chemical washing, during food processing, **229**: 101
Chemical-physical properties, ammonium polyphosphate (table), **222**: 15
Chemical-physical properties, HFFRs, **222**: 5
Chemical-physical properties, magnesium hydroxide (table), **222**: 13
Chemical-physical properties, phthalates, **224**: 41
Chemistry, clomazone, **229**: 36
Chemistry, diazinon, **223**: 107ff.
Chemistry, methomyl, **222**: 95
Chemodynamics, chlomazone, **229**: 37
Chemodynamics, methomyl, **222**: 95
Chicken (domestic), PCB toxicity, **230**: 64
Chicken toxicity, boron, **225**: 60, 67
Children, of addicted-substance mothers, **227**: 58
Chile, Santiago air contamination, **223**: 1ff.
Chilean guidelines, air pollutant hazards (table), **223**: 16
China, Hg as environmental contaminant, **223**: 54
China, TRVs for avian wildlife, **223**: 53 ff.
Chinese aquatic birds, PCB TRVs, **230**: 59 ff.
Chinese aquatic species hazard quotients, metals-metalloids (table), **230**: 50
Chinese aquatic species, risk assessment, **230**: 49
Chinese bird species, characterized (table), **223**: 56
Chinese bird species, methyl mercury levels (diag.), **223**: 73
Chinese bird species, most vulnerable to xenobiotics, **223**: 55
Chinese fish residues, heavy metals, **230**: 126
Chinese fish residues, methyl mercury (diag.), **223**: 70
Chinese pollutants, metals & metalloids, **230**: 36
Chinese residues, E2 risk assessment, **228**: 44
Chinese water quality criteria (WQC), metals & metalloids, **230**: 35 ff.
Chloramination, drinking water history, **222**: 114

Chlorinated water reactions, methomyl (diag.), **222**: 99
Chlorination, origins in drinking water, **222**: 112, 113
Chlorophyll effects, cadmium, **229**: 59
Chromate phytotoxicity, carbon amendment effect (illus.), **225**: 34
Chromate soil reduction, organic amendment effect (diag.), **225**: 33
Chromium characteristics, profile, **225**: 6
Chromium immobilization, in carbon amended soils (illus.), **225**: 38
Chromium in soil, uses (table), **225**: 5
Chromium soil metabolism, influencing factors (diag.), **225**: 24
Chromosomal aberrations, human bladder cancer role, **224**: 81
Chronic toxicity to aquatic species, diazinon, **223**: 130
Chronic toxicity to freshwater aquatic species, diazinon (table), **223**: 132
Classification scheme, HFFRs for bioaccumulation (table), **222**: 9
Classification scheme, HFFRs for persistence (table), **222**: 9
Classification scheme, HFFRs for toxicity (table), **222**: 9
Clastogen, ptaquiloside, **224**: 57
Clay effects, soil sorption processes, **221**: 20
Climate effects, soil column leaching, **221**: 14
Climate effects, zirconium release, **221**: 112
Climatic factors, influence on adult asthma, **226**: 48
Clinical profile, EBH in cattle (table), **224**: 56
Clinical symptoms, of EBH in cattle, **224**: 55
Clomazone toxicity, aquatic species, **229**: 44
Clomazone toxicity, to non-target species, **229**: 45
Clomazone, air & volatilization, **229**: 39
Clomazone, biotic processes, **229**: 40
Clomazone, chemistry & physicochemical properties, **229**: 36
Clomazone, description, **229**: 36
Clomazone, environmental chemodynamics, **229**: 37
Clomazone, environmental degradation, **229**: 39
Clomazone, environmental fate & toxicology, **229**: 35 ff.
Clomazone, metabolic pathway in soybeans (diag.), **229**: 44
Clomazone, microbial breakdown pathway (diag.), **229**: 41
Clomazone, photolysis & hydrolysis, **229**: 39

Clomazone, photolytic degradation (diag.), **229**: 40
Clomazone, physicochemical properties (table), **229**: 37
Clomazone, plant effects, **229**: 43
Clomazone, plant mode-of-action, **229**: 42
Clomazone, soil interactions, **229**: 37
Clomazone, structure (illus.), **229**: 36
Clomazone, toxicology, **229**: 42
Clomazone, water behavior, **229**: 38
CO, air contaminant, **223**: 5
Coal & biomass fuel exposure, adult asthma cause, **226**: 39
Coal deposits, acid mine drainage (table), **226**: 3
Coastal sediment levels, mercury, **226**: 68
Colloids, pesticide transport effects, **221**: 7
Combustion products, and adult asthma, **226**: 37
Common loon, methyl mercury toxicity, **223**: 63
Community effects, pollution, **227**: 79
Concentration effects, nanoparticles, **230**: 90
Consumption by users, polymers (table), **227**: 6
Consumption proportion, polymer types, **227**: 6
Consumption, PBMs, **227**: 3
Contaminants present, e-waste, **229**: 24
Contaminants present, in e-waste (table), **229**: 24
Contaminated soil remediation, with biochar, **228**: 83 ff.
Contaminated soil, biochar remediation (diag.), **228**: 85
Contamination by chemicals, drinking water guidelines (table), **222**: 116
Control methods for fungi, in water, **228**: 131
Control strategies, acid mine drainage, **226**: 4
Controlling EBH, approaches, **224**: 85
Copper accumulation levels, in land snails (table), **225**: 107
Copper bioaccumulation values, land snails (table), **225**: 101
Copper bioaccumulation, by land snails, **225**: 111
Copper biomarker in land snails, lysosomal stability, **225**: 117
Copper biomarker in land snails, membrane integrity, **225**: 117
Copper biomarker in land snails, metallothionein induction, **225**: 118
Copper exposure level, land snails, **225**: 99
Copper exposure routes, organisms, **225**: 97
Copper exposure sources, land snails, **225**: 99
Copper forms, toxicity & tolerance (table), **225**: 101
Copper in food, land snail exposure, **225**: 105

Copper mining, mercury release, **226**: 71
Copper nanoparticles, described, **230**: 99
Copper pollution biomarker, land snail oxidative stress, **225**: 119
Copper pollution monitoring, land snails (Pulmonata), **225**: 95 ff.
Copper pollution, land snail biomarkers, **225**: 113
Copper toxicity, among species (table), **225**: 98
Copper toxicity, expression methods, **225**: 100
Copper transfer, in soil-plant systems, **225**: 108
Copper uptake in land snails, shell vs. soft tissues, **225**: 113
Copper, air exposure, **225**: 100
Copper, as an environmental pollutant, **225**: 97
Copper, field exposure, **225**: 105
Copper, food chain movement, **225**: 105
Copper, land snail metabolic regulation, **225**: 108
Copper, plant levels, **225**: 106
Copper, soil exposure, **225**: 103
Criteria calculation, for sediment water quality, **224**: 131
Criteria setting for Chinese birds, toxicity data selection, **223**: 56
Critical study approach, in risk assessment, **230**: 73
Crop use in California, fenpropathrin (diag.), **225**: 80
Cu biomagnification, in aquatic ecosystems, **226**: 101 ff.
**D**aphnids, silver nanoparticle uptake, **223**: 96
Daphnids, silver toxicity, **223**: 93
Data analysis, setting sediment quality criteria for pesticides, **224**: 122
Data sources, for deriving sediment quality criteria, **224**: 114
DDT contamination, Brazilian coastal waters, **228**: 8
DDT residues, Guiana dolphins, **228**: 6
DDT residues, in small cetaceans (diag.), **228**: 8
DE-71 (PBDE isomer) NOAEL & LOAEL, American kestrels (table), **229**: 125
DE-71 NOAEL & LOAEL, mink (table), **229**: 119
DE-71 toxicity thresholds (diag.), **229**: 127
DE-71 toxicity, to birds, **229**: 121
Death rate in Americas, from air pollution, **223**: 3
Decomposition, phthalates, **224**: 43
Degradates of PBMs, potential effects (diag.), **227**: 25

Degradation half-life, fenpropathrin (table), **225**: 82
Degradation in the environment, PBMs, **227**: 14
Degradation of PBMs, abiotic processes, **227**: 15
Degradation pathways, fenpropathrin (diag.), **225**: 86
Degradation pathways, PBMs (diag.), **227**: 15
Degradation, methomyl, **222**: 98
Degradation, polymer-based materials, **227**: 1 ff.
Degradation, ptaquiloside, **224**: 77
Degradation-influencing factors, PBMs, **227**: 14
Demethylation in soils, trace elements (table), **225**: 18
Demethylation, of arsenic, **224**: 8
Demethylation, trace element detoxification, **225**: 16
Deoxynivalenol=vomitoxin, symptoms described, **228**: 109
Deriving TRC & TRV values, methodology, **223**: 57
Desulfurization, acid mine drainage control, **226**: 19
Detergent washing, during food processing, **229**: 101
Detoxification & transport, arsenic (diag.), **224**: 9
Detoxification mechanism, arsenic, **224**: 12
Diacetoxyscirpenol, a trichothecene mycotoxin, **228**: 109
Diagnosing bovine papillomavirus, polymerase chain reaction role, **224**: 85
Diazinon annual use, California urban vs. total (diag.), **223**: 111
Diazinon application methods, total use in California (diag.), **223**: 114
Diazinon effects, microbial populations, **223**: 121
Diazinon in California, air monitoring, **223**: 128
Diazinon insecticide, uses, **223**: 107
Diazinon persistence, aquatic environments, **223**: 125
Diazinon residues, California urban watersheds (table), **223**: 124
Diazinon residues, water, **223**: 122
Diazinon soil degradation, influencing factors, **223**: 120
Diazinon soil mobility, temperature effect, **223**: 118
Diazinon use volume, major California counties (table), **223**: 112
Diazinon water monitoring data, California agricultural counties (table), **223**: 124
Diazinon, acute & chronic toxicity to aquatic species, **223**: 130

Index

Diazinon, air degradation pathway (diag.), **223**: 127
Diazinon, aquatic life-benchmark and water quality criteria, **223**: 133
Diazinon, aquatic photolysis, **223**: 126
Diazinon, aquatic toxicity, **223**: 129
Diazinon, aquatic-life benchmark values (table), **223**: 133
Diazinon, behavior in air, **223**: 126
Diazinon, bioaccumulation, **223**: 132
Diazinon, California use profile, **223**: 107, 110
Diazinon, chemical structure (illus.), **223**: 108
Diazinon, chemistry & environmental fate, **223**: 107ff.
Diazinon, earthworm toxicity, **223**: 122
Diazinon, environmental fate, **223**: 114
Diazinon, ground- & surface-water residues, **223**: 122–3
Diazinon, groundwater detections, **223**: 108
Diazinon, mode of toxic action, **223**: 129
Diazinon, partition coefficient, **223**: 115
Diazinon, photolysis, **223**: 120
Diazinon, physicochemical properties (table), **223**: 109
Diazinon, physicochemical properties, **223**: 108
Diazinon, soil dissipation, **223**: 119
Diazinon, soil leaching, **223**: 117
Diazinon, soil microbial degradation, **223**: 119
Diazinon, soil mineralization, **223**: 118
Diazinon, soil photolysis pathway (diag.), **223**: 121
Diazinon, soil sorption, **223**: 115
Diazinon, soil-sediment degradation, **223**: 114
Diazinon, top crop uses in California (diag.), **223**: 113
Diazinon, urban vs. agricultural uses in California, **223**: 113
Diazinon, volatilization loss, **223**: 127
Diazinon, water quality criteria (table), **223**: 133
Dietary intake effects mapped, heavy metals (diag.), **226**: 108
Diphenyl ether pesticides, leaching characteristics, **221**: 73
Diphenyl ether pesticides, soil column leaching (table), **221**: 62–65
Disease etiology, EBH in cattle, **224**: 57
Disease in dolphins, environmental link, **228**: 13
Disease sentinels, Guiana dolphins, **228**: 1 ff.
Diseases, caused by fungi, **228**:129
Disinfectant residual loss, drinking water, **222**: 149
Disinfection by-products, formation in water **222**: 141

Disinfection by-products, formed in water (table), **222**: 142
Disinfection of drinking water, regulation, **222**: 115
Disinfection of water, successes & challenges, **222**: 111 ff.
Disintegration, PBMs, **227**: 21
Dispersion medium, nanoparticles, **230**: 89
Dissolved organic matter, pesticide transport effects, **221**: 7, 8
Distribution, mercury, **229**: 5
DNA damage by chemicals, implications, **227**: 83
Dolphin sentinels, ecotoxicity and emerging diseases, **228**: 1 ff.
Dolphin sentinels, marine ecotoxicity, **228**: 1 ff.
Double-crested cormorant, PCB toxicity, **230**: 68.
Drinking water contaminants, mycotoxins & aflatoxins, **228**:122
Drinking water contamination, by fungi, **228**:121 ff.
Drinking water contamination, sources of fungi, **228**:123
Drinking water disinfection, history, **222**: 112, 113
Drinking water disinfection, ozone, **222**: 115
Drinking water disinfection, recent advances, **222**: 135
Drinking water disinfection, regulation, **222**: 115
Drinking water disinfection, risk management concepts, **222**: 150, 153
Drinking water disinfection, successes & challenges, **222**: 111 ff.
Drinking water fungal contamination, common types, **228**:123
Drinking water guidelines, chemical contamination (table), **222**: 116
Drinking water history, chloramination, **222**: 114
Drinking water limits, chemical contaminants in EU (table), **222**: 123
Drinking water pathogens, of emerging concern, **222**: 129
Drinking water quality, Australia, **222**: 122
Drinking water quality, Canada, **222**: 120
Drinking water quality, microbial indicators, **222**: 128
Drinking water quality, South Africa, **222**: 117
Drinking water regulations, for US disinfectants (table), **222**: 119
Drinking water, biofilm formation & control, **222**: 144

Drinking water, biological pollutants, **228**:122
Drinking water, disinfectant residual loss, **222**: 149
Drinking water, emerging bacterial waterborne pathogens, **222**: 130
Drinking water, emerging fungal waterborne pathogens, **222**: 134
Drinking water, emerging viral & protozoan waterborne pathogens, **222**: 133
Drinking water, formation of disinfection by-products, **222**: 141
Drinking water, nitrification, **222**: 145
Drinking water, sources & types of fungi, **228**:123
Dry cover barrier, to prevent acidic mine effluents, **226**: 7
E2 (17-[beta]-estradiol), predicted no-effect concentration (PNEC) in Chinese waters, **228**: 29 ff.
E2 NOEC values, aquatic species (table), **228**: 38
E2 PNEC values, calculated via models (table), **228**: 39
E2 residues, in surface water, sediment and STP (sewage treatment plant) effluent (table), **228**: 45
E2 risk assessment, for Chinese concentrations, **228**: 44
E2 SSD (species sensitivity distribution), reproductive toxicity (diag.), **228**: 40
E2 toxicity, in aquatic organisms, **228**: 31
E2 toxicity, on aquatic species (table), **228**: 34–5
E2 toxicity, species sensitivity distribution (SSD) (diag.), **228**: 39
E2, amphibian toxicity, **228**: 33
E2, as an environmental contaminant, **228**: 30
E2, no observed effect concentration (NOEC) for reproduction, **228**: 37
E2, PNEC derivation method, **228**: 36
E2, rainbow trout toxicity, **228**: 34
Earthworm burrows, pesticide leaching effects, **221**: 33
Earthworm toxicity, diazinon, **223**: 122
Earthworms, soil macropore formation, **221**: 6
EBH (enzootic bovine hematuria), disease in cattle described, **224**: 54
*EBH* cause, *Pteridium aquilinum*, **224**: 54
EBH in cattle, caused by bracken & other ferns, **224**: 57
EBH in cattle, disease etiology, **224**: 57
EBH, clinical symptoms, **224**: 55
EBH, enzootic localities and incidence, **224**: 54
EBH, in India, **224**: 55
EBH, prevention & control, **224**: 85

EBH, symptoms & clinical character (table), **224**: 56
Ecosystem effects, pollution, **227**: 79
Ecotoxicity data needs, acute vs. chronic exposure, **224**: 120
Ecotoxicity data quality, deriving sediment quality criteria, **224**: 124
Ecotoxicity data quantity needed, deriving sediment quality criteria, **224**: 129
Ecotoxicity data, role in deriving sediment quality criteria, **224**: 120
Ecotoxicity sentinels, Guiana dolphins, **228**: 1 ff.
Ecotoxicity to aquatic species, fenpropathrin, **225**: 89
Ecotoxicity, fenpropathrin (table), **225**: 89
Ecotoxicity, HFFRs, **222**: 7
Ecotoxicology, fenpropathrin, **225**: 77 ff.
Ecotoxicology, fenpropathrin, **225**: 88
Ecotoxicology, use of sentinel organisms, **227**: 97
Effects of addictive substances, on mothers, fetuses & newborns (table), **227**: 60–3
Effects, PBM degradates (diag.), **227**: 25
Elasmobranch tissues, mercury residues (table), **226**: 80
Electrochemical cover, acid mine drainage control, **226**: 18
Electrochemical protection system, acid mine drainage control (illus.), **226**: 19
Elemental mercury, chemical behavior, **229**: 2
Elution medium effects, pesticide mobility (table), **221**: 37
Embryo development, bioamplification, **227**: 115
Emerging pathogens, in drinking water, **222**: 129
Empirical approaches, to developing sediment quality criteria, **224**: 107
Endangered species, implication for deriving sediment quality criteria, **224**: 157
Endocrine disrupting effects, of E2, **228**: 30
Endocrine disruptors, and cetaceans, **228**: 2
Endocrine-disrupting chemicals, description, **228**: 30
Endotoxin levels, in urban vs. rural areas, **226**: 42
Entanglement, PBMs, **227**: 24
Environmental aquatic toxicity, pesticides, **224**: 99
Environmental behavior, aluminum trihydroxide, **222**: 9
Environmental chemodynamics, clomazone, **229**: 37
Environmental concentrations, phthalates & BPA, **227**: 33
Environmental contaminant, E2, **228**: 30

Index 147

Environmental degradation, clomazone, **229**: 39
Environmental degradation, PBMs, **227**: 14
Environmental degradation, polymer-based materials, **227**: 1 ff.
Environmental distribution, arsenic, **224**: 4
Environmental effects, e-waste disposal, **229**: 19 ff.
Environmental effects, PBMs, **227**: 24
Environmental effects, polymer-based materials, **227**: 1 ff.
Environmental effects, ptaquiloside-induced enzootic bovine hematuria, **224**: 53 ff.
Environmental fate, arsenic, **224**: 4, 8–10
Environmental fate, clomazone, **229**: 35 ff.
Environmental fate, diazinon, **223**: 107ff., 114
Environmental fate, fenpropathrin, **225**: 77 ff.
Environmental fate, fenpropathrin, **225**: 82
Environmental fate, methomyl, **222**: 93 ff.
Environmental fate, polymer additives, **227**: 28
Environmental flow, nanowaste (diag.), **230**: 88
Environmental implication, polymer additives, **227**: 27
Environmental interactions, nanoparticles, **230**: 89
Environmental levels, natural arsenic (table), **224**: 3
Environmental management, arsenic cycle, **224**: 1 ff.
Environmental matrices, polymer degradation (table), **227**: 16–19
Environmental mobilization, arsenic, **224**: 10
Environmental movement & transport, fungi (diag.), **228**:130
Environmental occurrence, PBMs, **227**: 9
Environmental occurrence, polymer additives, **227**: 30
Environmental occurrence, polymer-based materials, **227**: 1 ff.
Environmental pollutants, metals & metalloids, **230**: 37
Environmental pollutants, polymer additives, **227**: 28
Environmental presence, HFFRs, **222**: 6
Environmental problem, acid mine drainage, **226**: 2
Environmental release, bulk PBMs, **227**: 7
Environmental release, nanoparticles, **230**: 87
Environmental residues detected, fenpropathrin, **225**: 79
Environmental residues, mercury, **226**: 66
Environmental residues, polymer additives (table), **227**: 30–2
Environmental sentinels, Guiana dolphins, **228**: 3

Environmental sentinels, marine mammals, **228**: 2
Environmental sink, PBM waste, **227**: 10
Environmental source, phthalates, **224**: 40
Environmental sources, cadmium, **229**: 52
Environmental stressors, disease link in dolphins, **228**: 13
Environmental tobacco smoke, adult asthma, **226**: 37
Enzootic bovine hematuria (EBH), ptaquiloside-induced, **224**: 53 ff.
Enzootic localities, EBH, **224**: 54
Enzymatic reduction, arsenic, **224**: 7
Enzyme inhibition, mercury, **229**: 10
Enzyme interactions, nanosilver toxicity, **223**: 98
Enzymes, boron effects, **225**: 63
Epididymal antioxidant system, bisphenol A effects, **228**: 70
Estrogen/androgen synthesis & action effects, BPA (table), **228**: 66
Estrogenic chemicals, described, **228**: 30
Ethyl mercury, thimerosal in vaccines, **229**: 4
EU (European Union) Directive, chemical contaminants in drinking water (table), **222**: 123
EU Directive, chemical contaminants in drinking water, **222**: 123
European trends, potable water disinfection, **222**: 127
E-waste contaminant effects, on aquatic organisms, **229**: 25
E-Waste contaminants (Korle Lagoon), vs. sediment quality guideline values (table), **229**: 25
E-Waste export, to developing countries, **229**: 20
E-Waste generated, globally, **229**: 20
E-Waste in Ghana, quantity imported (diag.), **229**: 21
E-Waste management, in Ghana, **229**: 21
E-Waste pollution issue, Ghana, **229**: 20
E-Waste quantity imported, into Ghana (diag.), **229**: 21
E-waste recycling, Agbogbloshie (Ghana) (illus.), **229**: 23
E-Waste recycling, in Ghana, **229**: 22
E-Waste water pollution, Agbogbloshie (Ghana) scrap yard, **229**: 22
E-Waste, aquatic environmental effects, **229**: 19 ff.
E-waste, contaminants present (table), **229**: 24
E-waste, contaminants present, **229**: 24
E-Waste, toxic chemicals present, **229**: 20
Exceedance probabilities, Santiago, Chile air pollutant data, **223**: 23

Exposure concentrations in a Chinese lake, metals-metalloids, **230**: 49
Exposure duration, setting sediment quality criteria, **224**: 132
Exposure frequency, setting sediment quality criteria, **224**: 132
Exposure magnitude, setting sediment quality criteria, **224**: 132
Exposure routes & sources, bisphenol A, **228**: 58
Eye exposure to radiation, cardiology staff, **222**: 82
Eyes, cardiology staff radiation dose (diags.), **222**: 80
**F**abricated nanoparticles, applications, **230**: 85
Fabricated nanoparticles, classes described, **230**: 85
Fabricated nanoparticles, commercial applications (table), **230**: 86
Fabricated nanoparticles, phytotoxicity, **230**: 83 ff.
Facilitated transport, pesticides in soil, **221**: 42
Fate in the body, mercury (diag.), **229**: 5
Fate in the environment, polymer additives, **227**: 28
Fatty acid composition, cadmium effects, **229**: 63
Fatty acids, acid mine drainage control, **226**: 17
Fenpropathrin ecotoxicity, aquatic species, **225**: 89
Fenpropathrin ecotoxicity, birds, **225**: 90
Fenpropathrin ecotoxicity, insects, **225**: 88
Fenpropathrin ecotoxicity, mammals, **225**: 91
Fenpropathrin, abiotic degradation, **225**: 85
Fenpropathrin, behavior in air, **225**: 82
Fenpropathrin, behavior in water, **225**: 83
Fenpropathrin, biotic degradation, **225**: 84
Fenpropathrin, California crop use (diag.), **225**: 80
Fenpropathrin, California use volume (2000–2010; diag.), **225**: 79
Fenpropathrin, degradation half-life (table), **225**: 82
Fenpropathrin, degradation pathways (diag.), **225**: 86
Fenpropathrin, description & characteristics, **225**: 78
Fenpropathrin, ecotoxicity (table), **225**: 89
Fenpropathrin, ecotoxicity, **225**: 77 ff.
Fenpropathrin, ecotoxicology, **225**: 88
Fenpropathrin, environmental fate, **225**: 77 ff.
Fenpropathrin, environmental fate, **225**: 82
Fenpropathrin, environmental residues detected, **225**: 79

Fenpropathrin, hydrolysis, **225**: 87
Fenpropathrin, mode of action, **225**: 80
Fenpropathrin, photolysis, **225**: 85
Fenpropathrin, physicochemical properties (table), **225**: 81
Fenpropathrin, physicochemical properties, **225**: 81
Fenpropathrin, plant metabolism, **225**: 85
Fenpropathrin, soil behavior, **225**: 83
Fenpropathrin, soil degradation, **225**: 84
Fenpropathrin, soil leaching behavior, **225**: 83
Fenpropathrin, structure (illus.), **225**: 78
Fenpropathrin, uses in the United States, **225**: 78
Fenpropathrin, weight used in California (diag.), **225**: 80
Fern carcinogen, ptaquiloside, **224**: 55
Fern content, ptaquiloside, **224**: 60
Fern fronds, ptaquiloside content, **224**: 60, 76
Fern-induced tumors, ptaquiloside adducts, **224**: 80
Fertilizer effects, pesticide mobility in soil, **221**: 43
Fetus effects, of addicted mothers (table), **227**: 60–3
Fetus specimen analysis, benefits (diag.), **227**: 68
Field data on tropic transfer, in aqueous environments, **226**: 106
Field exposure, copper, **225**: 105
Field leaching, pesticides, **221**: 16
Field studies, heavy metal accumulation results, **226**: 112
Field vs. laboratory studies, pesticide leaching, **221**: 53
Fingers, dose to cardiology staff (diag.), **222**: 79
Fish collection sites in Pakistan, heavy metal residues (illus.), **230**: 116
Fish heavy metal residues, China and Bangladesh, **230**: 126
Fish heavy metal residues, Iran and India, **230**: 127
Fish metal levels, spatial distribution (illus.), **230**: 117
Fish muscle residues, mercury (table), **226**: 82
Fish residues in China, methyl mercury (diag.), **223**: 70
Fish residues in Pakistan, heavy metals, **230**: 111 ff.
Fish residues, Khyber Pakhtunkhwa Province (Pakistan), **230**: 116
Fish residues, of heavy metals, **230**: 113
Fish residues, of mercury, **226**: 89
Fish residues, of PCBs, **230**: 77

Index

Fish residues, Punjab Province (Pakistan), **230**: 119
Fish residues, Sindh Province (Pakistan), **230**: 123
Fish toxicity mechanism, silver, **223**: 93
Fish toxicity, E2, **228**: 31
Fish, pollutant exposures and uptake, **230**: 112
Fishing gear entanglement, Guiana dolphins, **228**: 4
Flame retardant compounds, organophosphates, **222**: 22
Flame retardant, 9,10-Dihydro-9-oxa-10-phosphaphenanthrene-10-oxide, **222**: 42
Flame retardant, aluminum diethylphosphinate, **222**: 45
Flame retardant, bisphenol-A bis(diphenylphosphate), **222**: 37
Flame retardant, melamine polyphosphate, **222**: 48
Flame retardant, pentaerythritol, **222**: 51
Flame retardant, resorcinol bis(diphenylphosphate), **222**: 35
Flame retardant, triphenylphosphate, **222**: 22
Flame retardant, zinc borate, **222**: 16
Flame retardant, zinc hydroxystannate, **222**: 18
Flame retardant, zinc stannate, **222**: 21
Flame retardantcy requirements, organic materials, **222**: 2
Flame retardants, described, **222**: 2
Flyash amendment of soil, pesticide mobility effects, **221**: 40
Food (fruit & vegetable) processing methods, pyrethroid insecticide residue effects (table), **229**: 92–99
Food and water threat, fungal diseases, **228**:130
Food blanching, pesticide residue effect, **229**: 104
Food chain movement, copper, **225**: 105
Food chain risks, of polymer-based materials, **227**: 2
Food exposure to copper, land snails, **225**: 105
Food freezing, pesticide residue effect, **229**: 104
Food ingestion rates, avian species (table), **230**: 62
Food juicing, pesticide residue effect, **229**: 102
Food processing & pesticides, described, **229**: 90
Food processing effects, from boiling & cooking, **229**: 103
Food processing of vegetables & fruit, peeling, **229**: 102
Food processing, described, **229**: 91
Food processing, washing with chemicals or detergents, **229**: 101

Food processing, washing with salt solutions, **229**: 101
Food processing, washing with tap water, **229**: 91
Food washing effects, hydrophilic pesticide residues, **229**:100
Forehead, cardiology staff radiation dose (diags.), **222**: 80
Free amino acid effects, cadmium, **229**: 62
Freshwater fish residues, heavy metals, **230**: 113
Freshwater fish, heavy metal residues (table), **230**: 114
Freshwater species, metal & metalloid toxicity (table), **230**: 39
Freshwater trophic transfer factors, biokinetic heavy metal data (diag.), **226**: 110
Freshwater trophic transfer, heavy metals (diag.), **226**: 109
Fruit processing, pyrethroid insecticide residue effect, **229**: 89 ff.
Fumonisins, toxic fungal metabolites, **228**: 106
Fumonisins, toxic syndromes described, **228**: 106
Fungal allergen levels, in urban vs. rural areas, **226**: 42
Fungal diseases, threats, **228**:130
Fungal drinking water contaminants, common types, **228**:123
Fungal genera, disease causing, **228**:129
Fungal waterborne pathogens, in drinking water, **222**: 134
Fungal-produced toxins, mycotoxin risks, **228**:129
Fungi contamination, drinking water, **228**: 121 ff.
Fungi in drinking water, sources & types, **228**:123
Fungi in drinking water, sources, **228**:123
Fungi, boron effects, **225**: 62
Fungi, clustered genes & aflatoxin biosynthesis (diag.), **228**:128
Fungi, control in water sources, **228**:131
Fungi, environmental transport & movement (diag.), **228**:130
Fungi, fusarial toxin producers, **228**: 101 ff.
Fungi, identification methods, **228**:125
Fungi, isolation methods, **228**:124
Fungi, molecular identification methods, **228**:125
Fungi, universal PCR (polymerase chain reaction) primers (table), **228**:126
*Fusarial* toxins, prevention and control, **228**:110
Fusarial toxins, secondary metabolites of *Fusarium* fungi, **228**: 101 ff.

*Fusarium* fungal species, mycotoxins produced (table), **228**: 103
*Fusarium* fungi, fusarial toxins, **228**: 101 ff.
*Fusarium* fungi, mycotoxin producers, **228**: 102
Fusarium mycotoxins, structures (diag.), **228**:130
*Fusarium* toxins, cereal levels (table), **228**: 104
*Fusarium* toxins, host plants & growth influences, **228**: 104
*Fusarium*-produced mycotoxin, moniliformin, **228**: 107
*Fusarium*-produced mycotoxin, trichothecenes, **228**: 108
*Fusarium*-produced mycotoxins, *T*-2 and HT-2, **228**: 108
Gases, air contaminants, **223**: 5
Gaussian statistical modeling, atmospheric pollutants, **223**: 8
Gene mutations, Bracken fern, **224**: 79
Gene pool effects, xenobiotics, **227**: 82
Genetic damage monitoring, biomarker role, **227**: 84
Genetic diversity effects, biomarker detection, **227**: 85
Genetic diversity effects, metal pollution (table), **227**: 86–91
Genetic diversity effects, metals, **227**: 85
Genetic diversity in animals, metal effects, **227**: 92
Genetic diversity threats, metal pollution (diag.), **227**: 83
Genetic diversity, metal-pollution effects, **227**: 79 ff.
Genetic ecotoxicity, of metals, **227**: 80
Genetic ecotoxicity, sentinel organism use, **227**: 97
Genetic ecotoxicology, described, **227**: 80
Genetic effects, metal-exposed animals, **227**: 79 ff.
Genetic effects, metals, **227**: 79
Genetic effects, vs. natural genetic changes, **227**: 82
Genetic markers, to assess genetic variability, **227**: 95
Genetic material effects, cadmium, **229**: 63
Genetic structure response, selection forces, **227**: 80
Genetic structure, metal-pollution effects, **227**: 79 ff.
Genetic variability assessment, with genetic markers, **227**: 95
Genotoxic activity, ptaquiloside, **224**: 77
Genotoxicity, methylated As(III), **224**: 10
Genotoxin carcinogen, ptaquiloside, **224**: 57

Geographic distribution, Guiana dolphins (diag.), **228**: 4
Geographic factors, influence on adult asthma, **226**: 48
Ghana, e-waste disposal effects, **229**: 19 ff.
Ghana, e-waste management, **229**: 21
Ghana, e-waste pollution issue, **229**: 20
Ghana, e-waste recycling, **229**: 22
Giardiasis, in Guiana dolphins, **228**: 15
Global cycle, arsenic (diag.), **224**: 5
Global trends, potable water disinfection, **222**: 124
Gold nanoparticles, described, **230**: 101
Great egret, methyl mercury toxicity, **223**: 62
Great horned owl, PCB toxicity, **230**: 70
Groundwater contamination risk, mobility indices (table), **221**: 27
Groundwater residues, diazinon, **223**: 122
Groundwater ubiquity score (GUS), relative soil mobility (diag.), **221**: 30
Guiana dolphin blubber, organochlorine residues (table), **228**: 7
Guiana dolphin kidney & liver residues, Cd, Hg & Pb (table), **228**: 14
Guiana dolphin livers, metal accumulation, **228**: 12
Guiana dolphin residues, gender differences, **228**: 9
Guiana dolphin threat, lipophilic contaminants, **228**: 5
Guiana dolphin threats, environmental contamination, **228**: 4
Guiana dolphins, contaminants in prey, **228**: 5
Guiana dolphins, description (illus.), **228**: 3
Guiana dolphins, description, **228**: 3
Guiana dolphins, distribution (diag.), **228**: 4
Guiana dolphins, environmental sentinels, **228**: 3
Guiana dolphins, fishing gear entanglement, **228**: 4
Guiana dolphins, geographic distribution, **228**: 3
Guiana dolphins, hexachlorobenzene residues, **228**: 9
Guiana dolphins, hexachlorocyclohexane residues, **228**: 9
Guiana dolphins, human-related threats, **228**: 4
Guiana dolphins, lobomycosis & tattoo skin disease, **228**: 16
Guiana dolphins, marine ecosystem sentinels, **228**: 1 ff.
Guiana dolphins, PCB & DDT residues, **228**: 6
Guiana dolphins, PFCs, **228**: 10

Index 151

Guiana dolphins, toxoplasmosis & giardiasis, **228**: 15
Guidelines, air pollution in Santiago, Chile, **223**: 15
Guinea dolphins, human-induced injuries & marine debris, **228**: 18
Guinea dolphins, metal contamination, **228**: 11
Guinea dolphins, PBDEs, **228**: 10
Guinea dolphins, plastic debris effects, **228**: 19
Guinea dolphins, virus & helminth diseases, **228**: 17
GUS, relative pesticide mobility, **221**: 30
Halogenated pollutants, and cetaceans, **228**: 2
Halogen-free flame retardants (HFFRs), toxicity, **222**: 1 ff.
Halogen-free flame, persistence & bioaccumulation, **222**: 1 ff.
Halogens, for disinfecting drinking water, **222**: 139
Hands exposure to radiation, cardiology staff (diag.), **222**: 82
Hands, dose to cardiology staff (diag.), **222**: 79
Hazard quotients compared in Chinese aquatic species, metals-metalloids (table), **230**: 50
Hazard quotients, for Chinese aquatic species, **230**: 50
Hazardous concentration threshold (HC5) calculations, method comparisons (table), **230**: 42
Hazards of, acid mine drainage, **226**: 3
Hazards, polymer additives, **227**: 28
HC5 (hazardous concentration threshold) calculations, method comparisons (table), **230**: 42
HC5 value calculations, metals-metalloids (diag.), **230**: 44
HC5 values for Zn, analysis approach effects (diag.), **230**: 47
HC5 values, calculated via different methods (diag.), **230**: 44
Health effects from irradiation, cardiology staff, **222**: 85
Health effects from radiation, cardiology staff, **222**: 73 ff.
Health effects, air contaminants, **223**: 5
Health effects, air pollution, **223**: 2
Health effects, phthalates, **224**: 40
Healthcare access effects, asthma prevalence, **226**: 49
Heavy metal accumulation results, from field studies, **226**: 112
Heavy metal accumulation, Guiana dolphin livers, **228**: 12

Heavy metal binding, land snails, **225**: 97
Heavy metal biokinetic data, freshwater trophic transfer factors (diag.), **226**: 110
Heavy metal characteristics, described, **230**: 112
Heavy metal concentrations in fish, spatial distribution (illus.), **230**: 117
Heavy metal effects, on aquatic life, **229**: 26
Heavy metal induction, reactive oxygen species, **223**: 40
Heavy metal phytoremediation, rhizobacteria role, **223**: 33ff.
Heavy metal plant stress, PGPR (plant growth-promoting rhizobacteria) counter effect, **223**: 42
Heavy metal residues, in fish species, **230**: 113
Heavy metal residues, in freshwater fish, **230**: 113
Heavy metal residues, in marine & freshwater fish (table), **230**: 114
Heavy metal residues, in Pakistani fish, **230**: 111 ff.
Heavy metal residues, in Pakistani marine fish, **230**: 124
Heavy metal trophic transfer, biokinetic-derived marine data (diag.), **226**: 111
Heavy metals in fish, Bangladesh and China, **230**: 126
Heavy metals in fish, India and Iran, **230**: 127
Heavy metals in soils, uses (table), **225**: 5
Heavy metals, air contaminants, **223**: 7
Heavy metals, described, **229**: 52
Heavy metals, dietary intake effects mapped (diag.), **226**: 108
Heavy metals, dietary tropic transfer factors (diag.), **226**: 107
Heavy metals, do they biomagnify?, **226**: 101 ff.
Heavy metals, exposures & interactions, **223**: 34
Heavy metals, in Guiana dolphins, **228**: 11
Heavy metals, remediation, **223**: 35
Heavy metals, role in plants, **223**: 34
Heavy metals, soil bioavailability, **223**: 37
Heavy metals, toxic mechanisms in plants, **223**: 34
Heavy-metal remediation, PGPR role, **223**: 43
Helminth diseases, of Guinea dolphins, **228**: 17
Herpes virus resistance, boron in cattle, **225**: 65
Herring gull bioamplification factors, modeling (table), **227**: 131
Herring gull model, bioamplification simulation, **227**: 141
Herring gull simulation, bioamplification model parameters (table), **227**: 143–5
Herring gulls, bioamplification modeling (diags.), **227**: 130

Hexachlorobenzene residues, Guiana dolphins, **228**: 9
Hexachlorocyclohexane residues, Guiana dolphins, **228**: 9
HFFR (halogen-free flame retardant), properties, environmental character (table), **222**: 59
HFFRs, brominated flame retardant alternatives (table), **222**: 4
HFFRs, categories & uses, **222**: 3
HFFRs, characteristics, **222**: 5
HFFRs, classification & categories, **222**: 8
HFFRs, classification of persistence, toxicity & bioaccumulation (table), **222**: 9
HFFRs, data availability & consistency, **222**: 57
HFFRs, environmental behavior, **222**: 58
HFFRs, environmental presence & production volumes, **222**: 6
HFFRs, *in vitro* toxicity, **222**: 8
HFFRs, *in vivo* toxicity, **222**: 7
HFFRs, persistence & bioaccumulation, **222**: 6
HFFRs, physical-chemical properties, **222**: 5
HFFRs, REACH system data, **222**: 5
HFFRs, toxicity & ecotoxicity, **222**: 7
Hg adsorption variables, stack emissions, **230**: 28
Hg emissions removal, results summary (table), **230**: 6–25
Hg emissions to air, removal methods, **230**: 2
Hg removal efficiency, by S-impregnated activated carbon, **230**: 26
Hg removal efficiency, virgin- vs. S-impregnated-activated carbons, **230**: 26
Hg removal from airstacks, results summary (table), **230**: 6–25
Hg removal, by S-impregnated activated carbon, **230**: 4
Hg residues, in Guiana dolphin kidney & liver (table), **228**: 14
$Hg^0$ adsorption effects, sulfur to carbon ratio, **230**: 28
$Hg^0$ adsorption, inlet concentration effects, **230**: 28
$Hg^0$ adsorption, sulfur-content effects, **230**: 28
$Hg^0$ adsorption, temperature effects, **230**: 27
Hibernation effects, bioamplification, **227**: 121
Host plants, *Fusarium* toxins, **228**: 104
Human activities, mercury release, **226**: 66
Human bladder cancer, chromosomal aberration role, **224**: 81
Human consumed seafood, mercury levels, **226**: 89
Human effects, mercury, **226**: 89

Human effects, ptaquiloside-induced enzootic bovine hematuria, **224**: 53 ff.
Human exposure pathways, adult asthma causes (diag.), **226**: 35
Human exposure risks, from mining activities, **226**: 90
Human exposure sources, mercury, **229**: 3
Human exposure, to mercury, **229**: 3
Human fatality, boron toxicity, **225**: 67
Human food chain, flow of ptaquiloside (diag.), **224**: 75
Human food chain, ptaquiloside entry points (diag.), **224**: 74
Human health effects, ptaquiloside, **224**: 73
Human health effects, relation to statistical information, **223**: 24
Human health effects, Santiago, Chile air pollution, **223**: 15
Human male effects, bisphenol A *in utero* exposure, **228**: 60–2
Human morbidity, urban vs. rural affects, **226**: 34
Human per capita consumption, polymers (table), **227**: 6
Human poisoning events, methyl mercury, **229**: 4
Human residues, bisphenol A, **228**: 59
Human-induced injuries, Guinea dolphins, **228**: 18
Human-related threats, to Guiana dolphins, **228**: 4
Humans, boron intake & sources, **225**: 61
Humans, boron intake effects, **225**: 61
Humans, boron toxicity, **225**: 66
Humic acid amendment of soil, pesticide mobility effects, **221**: 40
Humic acid, acid mine drainage control, **226**: 15
Hydrolysis, clomazone, **229**: 39
Hydrolysis, fenpropathrin, **225**: 87
Hydrolysis, methomyl, **222**: 98
Hydrolysis, phthalate esters, **224**: 44
Hydrolysis, phthalates transformation (diag.), **224**: 45
Hydrolytic degradation, PBMs, **227**: 21
Hyperaccumulator plants, in phytoremediaton, **223**: 37
Hyperaccumulator plants, metal bioavailability effect, **223**: 39
Hypothalamic-pituitary-testicular axis effects, of bisphenol A, **228**: 62
Hysteresis effects, pesticide absorption-desorption, **221**: 21

**I**mmunohistochemical expression, tumor biomarkers, **224**: 80

Index 153

Implications, bioamplification process, **227**: 133
Impregnation time, Hg adsorption from stack emissions, **230**: 30
*In utero* biomonitoring, addictive substances, **227**: 63
*In vitro* toxicity, silver, **223**: 83, 92
India, EBH disease in cattle, **224**: 55
Indian fish residues, heavy metals, **230**: 127
Indoor air pollution, people effects, **226**: 34
Industrial sources, PBM pollution, **227**: 8
Infants, addictive substance effects, **227**: 55 ff.
Ingestion, PBMs, **227**: 24
Inorganic coatings, controlling acid mine drainage, **226**: 9
Inorganic flame retardant, aluminum trihydroxide character (table), **222**: 11
Inorganic flame retardant, aluminum trihydroxide, **222**: 9
Inorganic flame retardant, ammonium polyphosphate, **222**: 13
Inorganic flame retardant, magnesium hydroxide, **222**: 12
Inorganic flame retardants, described, **222**: 9
Inorganic flame retardants, synergists, **222**: 9
Inorganic mercury, in breast milk, **229**: 6
Insecticide described, methomyl, **222**: 94
Insecticide residue effects, from vegetable & fruit processing, **229**: 89 ff.
Insects, fenpropathrin ecotoxicity, **225**: 88
Insects, methomyl toxicity, **222**: 102
International standards, maximum concentrations for air pollutants (table), **223**: 16
Interventional cardiology staff, radiation doses, **222**: 78
Interventional cardiology staff, radiation exposure, **222**: 74
Intumescent flame retardant, pentaerythritol, **222**: 51
Inversion effects, Santiago, Chile air pollution, **223**: 14
Invertebrates, Mexican mercury residues, **226**: 77
Iranian fish residues, heavy metals, **230**: 127
Iraqi poisoning event, methyl mercury, **229**: 4
Iron oxidizing bacteria, mineral oxidation, **226**: 2
Isolation methods, fungi, **228**:124
Juvenile development, bioamplification, **227**: 115
Kawasaki disease, mercury link, **229**: 6
Kestrel (American), PCB toxicity, **230**: 70
Khyber Pakhtunkhwa Province (Pakistan), fish residues, **230**: 116

Laboratory vs. field studies, pesticide leaching, **221**: 53
Land snail accumulation, copper (table), **225**: 107
Land snail biomarker, lysosomal stability, **225**: 117
Land snail biomarker, membrane integrity, **225**: 117
Land snail biomarker, metallothionein induction, **225**: 118
Land snail biomarkers, copper pollution, **225**: 113
Land snail biomarkers, dependencies (table), **225**: 115
Land snail biomarkers, histopathology & ultrastructure, **225**: 114
Land snail exposure, copper in soil, **225**: 103
Land snail metabolic regulation, copper, **225**: 108
Land snail oxidative stress, biomarker of copper pollution, **225**: 119
Land snail tolerance, chemical elements, **225**: 97
Land snail use, terrestrial ecosystem monitoring, **225**: 95 ff.
Land snails, biological attributes, **225**: 96
Land snails, copper bioaccumulation, **225**: 111
Land snails, copper exposure levels, **225**: 99
Land snails, copper exposure sources, **225**: 99
Land snails, copper pollution monitoring, **225**: 95 ff.
Land snails, copper soil pollution, **225**: 103
Land snails, field exposure to copper, **225**: 106
Land waste, macro PBMs, **227**: 10
Land waste, micro PBMs, **227**: 13
Land-based pollution, by PBMs, **227**: 7
Landfills, PBM disposal, **227**: 8
Landfills, phthalate behavior, **224**: 39 ff.
Landfills, source of phthalate esters, **224**: 42
Land-snail bioaccumulation values, copper (table), **225**: 101
Land-snail exposure, copper in food, **225**: 105
Leaching behavior, pesticides, **221**: 2
Leaching characteristics, pesticide classes, **221**: 73
Leaching potential, fenpropathrin, **225**: 83
Leaching study types, relative reliability, **221**: 53
Lead toxicity, to aquatic organisms, **229**: 27
Lipid loss, bioamplification implication, **227**: 113
Lipid reserve management, animal energetics, **227**: 114
Lipid use, acid mine drainage control, **226**: 15

Lobomycosis, in Guiana dolphins, **228**: 16
Log-normal distribution, fitting air pollutant data, **223**: 12
Lysimeter (outdoor) results, pesticides (diag.), **221**: 72
Lysimeter advantages, pesticide mobility measurement, **221**: 16
Lysimeter studies, pesticide loss rate, **221**: 15
Lysimeter studies, use of radiotracers, **221**: 15
Lysimeters (outdoor), measuring pesticide mobility, **221**: 14
Lysosomal stability, land snail biomarker, **225**: 117
**M**acro PBM waste, ocean, **227**: 9
Macro PBM waste, on shore and land, **227**: 10
Macrophyte mercury levels, Mexico (table), **226**: 88
Macropore disruption, soil tillage, **221**: 31
Macropores in soil, definition, **221**: 4
Macropores, formation in soils, **221**: 5
Magnesium hydroxide, chemical & environmental character, **222**: 12
Magnesium hydroxide, toxicity & physical-chemical properties (table), **222**: 13
Magnetic nanoparticles, described, **230**: 101
Male reproductive effects, bisphenol A, **228**: 57 ff.
Male reproductive effects, BPA (table), **228**: 63
Male reproductive function, BPA action sites (diag.), **228**: 71
Mallard, methyl mercury toxicity, **223**: 58
Mammal residues in Mexico, mercury, **226**: 84
Mammal tissue residues, mercury (table), **226**: 85
Mammalian ecotoxicity, fenpropathrin, **225**: 91
Mammalian toxicity, methomyl, **222**: 104
Mammals, PBDE TRVs (table), **229**: 127
Mammals, pollution assessment, **227**: 98
Manufacturing process, PBMs, **227**: 3
Marine accumulator species, biokinetic trophic transfer data (diag.), **226**: 112
Marine biokinetic data, heavy metal trophic transfer (diag.), **226**: 111
Marine debris component, PBMs (table), **227**: 11
Marine debris, biodiversity effects, **228**: 19
Marine debris, Guinea dolphin effects, **228**: 18
Marine fish residues in Pakistan, heavy metals, **230**: 124
Marine fish toxicity, of E2, **228**: 31
Marine fish, heavy metal residues (table), **230**: 114
Marine habitat, PBM sink, **227**: 10
Marine mammal threat, pathogens, **228**: 2

Marine mammals, as environmental sentinels, **228**: 2
Marine mammals, persistent organic contaminants, **228**: 5
Marine mammals, toxoplasmosis affliction, **228**: 15
Marine pathogens, marine mammal threat, **228**: 2
Marine species, pollutant threats, **228**: 2
Maritime anti-dumping rules, PBMs, **227**: 7
Maternal absorption & transport, addictive substances, **227**: 66
Maternal effects, of addictive substances (table), **227**: 60–3
Maternal specimen analysis, benefits (diag.), **227**: 68
Maternal transport to child, of xenobiotics (diag.), **227**: 65
Mathematical modeling, pesticide transport in soil, **221**: 50
Maximum residual disinfectant levels, USEPA (table), **222**: 118
Measured exposure concentrations, metals-metalloids in a Chinese Lake (table), **230**: 49
Measurement in organisms, bioamplification, **227**: 112
Mechanical disintegration, PBMs, **227**: 21
Mechanism, of arsenic detoxification (diag.), **224**: 11
Mechanistic approach, to developing sediment quality guidelines, **224**: 107
MeHg toxicity, effects described, **229**: 7
MeHg, selenium effects, **229**: 8
MeHg, thiol interactions, **229**: 7
Melamine polyphosphate, flame retardant, **222**: 48
Melamine polyphosphate, persistence & bioaccumulation, **222**: 50, 51
Melamine polyphosphate, physical-chemical & toxic properties (table), **222**: 49
Membrane integrity, land snail biomarker, **225**: 117
Membrane technology, drinking water disinfection, **222**: 137
Mercury characteristics, profile, **225**: 6
Mercury chelators, structures (illus.), **229**: 9
Mercury contamination, in small cetacean liver (diag.), **228**: 12
Mercury cycle, complexity, **226**: 66
Mercury exposure, humans, **229**: 3
Mercury fish residues, muscle tissue (table), **226**: 82
Mercury forms, human exposure, **229**: 3

Index 155

Mercury in soil, uses (table), **225**: 5
Mercury levels in Mexico, molluscs (table), **226**: 78
Mercury levels in Mexico, shrimp (table), **226**: 78
Mercury levels, in human consumed seafood, **226**: 89
Mercury levels, in Mexican coastal waters (table), **226**: 76
Mercury levels, macrophytes & vestimentiferan tube worms (table), **226**: 88
Mercury levels, Mexican coastal sediments (table), **226**: 69
Mercury link, amyotrophic lateral sclerosis, **229**: 6, 10
Mercury link, Kawasaki disease, **229**: 6
Mercury metabolism, influencing factors, **225**: 22
Mercury nephrotoxic effects, on humans, **226**: 90
Mercury release levels, to the atmosphere, **226**: 67
Mercury removal, by sulfur-impregnated activated carbon, **230**: 1 ff.
Mercury removal, stack emissions, **230**: 1 ff.
Mercury residues, aquatic biota in Mexico, **226**: 77
Mercury residues, in consumable fish, **226**: 89
Mercury residues, in reptiles, birds and mammals, **226**: 84
Mercury residues, in the aquatic environment, **226**: 68
Mercury residues, Mexican coastal waters, **226**: 75
Mercury residues, Mexican invertebrates, **226**: 77
Mercury residues, Mexican vertebrates, **226**: 79
Mercury tissue levels, birds, reptiles & mammals (table), **226**: 85
Mercury tissue residues, in Mexican elasmobranchs (table), **226**: 80
Mercury toxicity, metallothionein effects, **229**: 8
Mercury toxicity, mitochondrial effects, **229**: 8
Mercury toxicity, molecular mechanism, **229**: 7
Mercury toxicity, neurodegenerative effects, **229**: 1 ff.
Mercury toxicity, reactive oxygen species, **229**: 8
Mercury toxicity, symptoms, **229**: 6
Mercury vapor, toxicity, **229**: 6
Mercury, absorption and distribution, **229**: 5
Mercury, Alzheimer's disease, **229**: 10
Mercury, and neurodegenerative diseases, **229**: 10
Mercury, anthropogenic releases, **226**: 66
Mercury, as Chinese environmental contaminant, **223**: 54
Mercury, as global environmental contaminant, **223**: 54
Mercury, as worldwide contaminant, **226**: 65
Mercury, autoimmune effects, **229**: 6
Mercury, biogeochemical cycle, **226**: 66
Mercury, chemical forms and properties, **229**: 2
Mercury, coastal sediment levels, **226**: 68–75
Mercury, effects described, **230**: 2
Mercury, environmental residues, **226**: 66
Mercury, enzyme inhibition, **229**: 10
Mercury, fate in the body (diag.), **229**: 5
Mercury, from copper mining, **226**: 71
Mercury, hazardous air pollutant, **230**: 2
Mercury, human effects, **226**: 89
Mercury, in Mexico, **226**: 65 ff.
Mercury, in the atmosphere, **226**: 67
Mercury, neuronal degeneration, **229**: 7
Mercury, organic vs. reactive forms (diag.), **225**: 22
Mercury, releases to environment, **230**: 2
Mercury, sources, **226**: 66
Mercury, total gaseous levels released (table), **226**: 68
Mercury, toxicity, **229**: 5
Metabolic pathway in soybeans, chlomazone (diag.), **229**: 44
Metabolic regulation in land snails, copper, **225**: 108
Metabolism, addictive substances, **227**: 65
Metabolites, abused substance biomonitoring (table), **227**: 68–9
Metal & metalloid toxicity, freshwater species (table), **230**: 39
Metal adsorption, biochar, **228**: 86
Metal contamination, Guiana dolphins, **228**: 11
Metal disinfectants, drinking water disinfection, **222**: 138
Metal effects, genetic diversity in animals, **227**: 92
Metal effects, genetic diversity, **227**: 85
Metal pollutants, and cetaceans, **228**: 2
Metal pollutants, genetic effects, **227**: 79
Metal pollutants, sources, **223**: 33
Metal pollution effects, animal genetic diversity (table), **227**: 86–91
Metal pollution effects, genetic diversity & structure, **227**: 79 ff.

Metal pollution, genetic diversity threats (diag.), **227**: 82
Metal species in soil, uses (table), **225**: 5
Metal stress, population genetic response, **227**: 84
Metal toxicity, population genetic response, **227**: 81
Metal uptake in plants, cadmium effects, **229**: 64
Metal-binding properties, metallothionein proteins, **223**: 39
Metal-exposed animals, genetic effects, **227**: 79 ff.
Metallic elements toxicity, species sensitivity, **230**: 46
Metalloid contamination, bioremediation, **225**: 38
Metalloids, setting WQC in China, **230**: 35 ff.
Metallothionein effects, mercury toxicity, **229**: 8
Metallothionein induction, land snail biomarker, **225**: 118
Metallothionein proteins, metal-binding properties, **223**: 39
Metal-metalloid concentrations, in Chinese surface waters, **230**: 39
Metal-metalloid toxicity, SSD bootstrap regression (diag.), **230**: 48
Metals & metalloids, Chinese pollutants, **230**: 36
Metals and metalloids, environmental pollutants, **230**: 37
Metals, genetic ecotoxicity, **227**: 80
Metals, plant detoxification mechanisms, **223**: 39
Metals, plant uptake and transport, **223**: 38
Metals, setting WQC in China, **230**: 35 ff.
Metals, soil bioavailability, **223**: 37
Metals-metalloids in a Chinese Lake, sampling sites (illus.), **230**: 40
Metals-metalloids, Chinese lake exposure concentrations, **230**: 49
Metals-metalloids, correlation coefficients (table), **230**: 50
Metals-metalloids, hazard quotients compared in aquatic species (table), **230**: 50
Metals-metalloids, hazard quotients for Chinese aquatic species, **230**: 50
Metals-metalloids, HC5 value calculations (diag.), **230**: 44
Metals-metalloids, measured exposure concentrations in a Chinese lake (table), **230**: 49
Metals-metalloids, risk assessment procedure, **230**: 43

Metamorphosis effects, bioamplification, **227**: 119
Method effects, pesticide dissipation rate, **221**: 24
Methodology components, sediment quality criteria, **224**: 101
Methods discussed, for WQC setting, **230**: 51
Methomyl degradation, hydrolysis & photolysis, **222**: 98
Methomyl toxicity, aquatic organisms (table), **222**: 104
Methomyl, aquatic species toxicity, **222**: 103
Methomyl, behavior in air, **222**: 98
Methomyl, biotic degradation, **222**: 101
Methomyl, bird & mammalian toxicity, **222**: 104
Methomyl, chemical structure (illus.), **222**: 94
Methomyl, chemistry & chemodynamics, **222**: 95
Methomyl, chlorinated water reactions (diag.), **222**: 99
Methomyl, described, **222**: 94
Methomyl, environmental degradation, **222**: 98
Methomyl, environmental fate & toxicity, **222**: 93 ff.
Methomyl, insect toxicity, **222**: 102
Methomyl, photocatalytic breakdown (diag.), **222**: 100
Methomyl, physicochemical properties (table), **222**: 95
Methomyl, soil behavior, **222**: 95
Methomyl, toxicity & mode of action, **222**: 102
Methomyl, toxicity, **222**: 94
Methomyl, water behavior, **222**: 96
Methyl mercury avian toxicity, species sensitivity distribution (diag.), **223**: 68
Methyl mercury in China, bird protection criteria, **223**: 53 ff.
Methyl mercury levels, avian species in China (diag.), **223**: 73
Methyl mercury toxicity, common Chinese wildfowl species, **223**: 58–64
Methyl mercury, avian species toxicity thresholds (diag.), **223**: 65
Methyl mercury, avian wildlife TRV uncertainty factors (table), **223**: 66
Methyl mercury, avian wildlife TRVs & TRCs (table), **223**: 67
Methyl mercury, bird reproductive effects, **223**: 62
Methyl mercury, bird toxicity, **223**: 58
Methyl mercury, fish concentrations in China (diag.), **223**: 70
Methyl mercury, human poisoning events, **229**: 4

Index 157

Methyl mercury, Minamata disease, **225**: 3
Methyl mercury, nature & origin, **229**: 2
Methyl mercury, sub-chronic & chronic avian toxicity (table), **223**: 59–61
Methyl mercury, toxicity thresholds for avian species (table), **223**: 67
Methyl mercury, wildlife contaminant, **223**: 55
Methylarsines, arsenic metabolism, **224**: 9
Methylated As(III), genotoxicity, **224**: 10
Methylation & demethylation, role in arsenic transport, **224**: 9
Methylation in soils, trace elements (table), **225**: 18
Methylation, of arsenic, **224**: 8
Methylation, trace element detoxification, **225**: 16
Methylmercury, soil degradation, **225**: 19
Mexican bird tissue residues, mercury (table), **226**: 85
Mexican coastal areas, mercury residues, **226**: 65 ff.
Mexican coastal sediment levels, mercury (table), **226**: 69
Mexican coastal waters, mercury levels (table), **226**: 76
Mexican coastal waters, mercury residues, **226**: 75
Mexican elasmobranchs, mercury tissue residues (table), **226**: 80
Mexican macrophytes & vestimentiferan tube worms, mercury levels (table), **226**: 88
Mexican mammal residues, mercury (table), **226**: 85
Mexican mercury contamination, invertebrates, **226**: 77
Mexican mercury contamination, vertebrates, **226**: 79
Mexican mollusks, mercury residues (table), **226**: 78
Mexican reptile residues, mercury (table), **226**: 85
Mexican reptiles, birds and mammals, mercury residues, **226**: 84
Mexican shrimp, mercury residues (table), **226**: 78
Mexican teleost fish, mercury residues (table), **226**: 82
Mexico, mercury residues in aquatic biota, **226**: 77
Mexico, release of gaseous mercury (table), **226**: 68
Micro PBM waste, ocean, **227**: 13
Micro PBM waste, shores and land, **227**: 13

Microbe involvement, chemical degradation, **221**: 25
Microbes, resistant to arsenic (table), **224**: 6
Microbes, toxin sensing, **224**: 26
Microbial activity, trace element metabolism, **225**: 21
Microbial aspects, trace element immobilization (diag.), **225**: 36
Microbial bioaccumulation, trace elements, **225**: 8
Microbial biomethylation, of trace elements in soil, **225**: 17
Microbial breakdown pathway, clomazone (diag.), **229**: 41
Microbial contaminants, Canadian drinking water guidelines (table), **222**: 121
Microbial detoxification & transport, arsenic (diag.), **224**: 9
Microbial indicators, drinking water quality, **222**: 128
Microbial interactions in soil, with trace elements, **225**: 11
Microbial interactions, arsenic cycle, **224**: 1 ff.
Microbial interactions, trace elements, **225**: 2
Microbial oxidation, arsenite, **224**: 7
Microbial population effects, diazinon, **223**: 121
Microbial redox reactions, trace elements, **225**: 12
Microbial resistance, arsenic, **224**: 6
Microbial risk assessment, in water disinfection, **222**: 153
Microbial sensing, arsenic, **224**: 13
Microbial transformation processes, trace elements (diag.), **225**: 8
Microbial transformation processes, trace elements, **225**: 20
Microbial transformation, in arsenic cycle, **224**: 6
Microbial transformation, processes, **225**: 7
Microbial transformation, trace elements in soil, **225**: 1ff.
Microbial transformation, trace elements, **225**: 3
Microbial uptake, arsenic, **224**: 10
Microbiological standards, for drinking water (table), **222**: 117
Microencapsulation, controlling acid mine drainage, **226**: 9
Microplastics, oceanic waste, **227**: 13
Migration effects, on bioamplification, **227**: 122
Minamata disease, mercury poisoning event, **229**: 4
Minamata disease, methyl mercury, **225**: 3
Mineral effects, on biochar soil remediation performance, **228**: 91

Mineral oxidation role, phosphate coatings, **226**: 10
Mineral oxidation, iron oxidizing bacterial, **226**: 2
Mineral oxidation, to form acid mine drainage, **226**: 2
Minerals, boron effects, **225**: 63
Minerals, boron regulatory role, **225**: 65
Minerals, linked to acid mine drainage (table), **226**: 3
Mining activities, human mercury exposure, **226**: 90
Minks, DE-71 (PBDE isomer), NOAEL & LOAEL (Table), **229**: 119
Mitigation, acid mine drainage, **226**: 1 ff.
Mitochondrial effects, mercury toxicity, **229**: 8
Mobility indices, groundwater contamination risk (table), **221**: 27
Mobility indices, pesticides, **221**: 26
Mobility of pesticides, soil column leaching studies, **221**: 52
Mobilization & demobilization, arsenic, **224**: 10
Mode of action for nanosilver, apoptosis via ROS, **223**: 87
Mode of action, fenpropathrin, **225**: 80
Mode of action, methomyl, **222**: 102
Mode of toxic action, diazinon, **223**: 129
Model simulations, bioamplification factors (table), **227**: 129, 131
Model, pesticide soil mobility (diag.), **221**: 46
Modeling approaches, bioamplification of POPs, **227**: 124
Modeling soil mobility, pesticides, **221**: 45
Modeling, pesticide transport in soil, **221**: 50
Modes of action possibilities, nanosilver (diag.), **223**: 88
Modes of action, nanosilver, **223**: 81 ff.
Molecular identification methods, fungi, **228**:125
Molecular mechanism, mercury toxicity, **229**: 7
Molecular tools, to assess mutagenicity, **227**: 84
Molluscs, mercury levels in Mexico (table), **226**: 78
Molybdenum nanoparticles, description, **230**: 100
Moniliformin, *Fusarium*-produced mycotoxin, **228**: 107
Morbidity in rural & urban areas, from adult asthma, **226**: 36
Multipathway exposures, in deriving sediment quality criteria, **224**: 134
Multiwalled carbon nanotubes (MWCNTs), description, **230**: 92
Mutagen, ptaquiloside, **224**: 57

Mutagenicity assessment, molecular & phylogenetic tools, **227**: 84
MWCNTs (multiwalled carbon nanotubes), description, **230**: 92
Mycotoxin producers, *Fusarium* fungi, **228**: 102
Mycotoxin risks, fungal-produced toxins, **228**:129
Mycotoxins produced, by selected fungal species (table), **228**: 103
Mycotoxins, analysis and detection, **228**:113
Mycotoxins, characteristics, **228**:130
Mycotoxins, described, **228**: 101
Mycotoxins, drinking water contaminant, **228**:122
Nanoparticle aggregation, toxicity effect, **223**: 94
Nanoparticle penetration, into plants, **230**: 91
Nanoparticle release, to the environment, **230**: 87
Nanoparticle risk assessment scheme, for plants (diag.), **230**: 103
Nanoparticle silver uptake, daphnids, **223**: 96
Nanoparticle toxic effects, a survey (table), **230**: 93–95
Nanoparticle toxicity role, silver ions, **223**: 97
Nanoparticle, factors affecting phytotoxicity, **230**: 88
Nanoparticle-plant interactions, mediating factors (diag.), **230**: 89
Nanoparticles described, copper- & cerium-based, **230**: 99
Nanoparticles described, gold-based, **230**: 101
Nanoparticles described, magnetic and other types, **230**: 101
Nanoparticles described, molybdenum-based, **230**: 100
Nanoparticles described, silver-based, **230**: 97
Nanoparticles described, titanium-based, **230**: 96
Nanoparticles described, zink-based, **230**: 96
Nanoparticles, and plants, **230**: 87
Nanoparticles, commercial applications (table), **230**: 86
Nanoparticles, concentration effects, **230**: 90
Nanoparticles, description, **223**: 81
Nanoparticles, environmental interactions, **230**: 89
Nanoparticles, natural & man-made (table), **230**: 84
Nanoparticles, particle size effects, **230**: 90
Nanoparticles, phytotoxic effects, **230**: 92
Nanoparticles, phytotoxic threats, **230**: 83 ff.
Nanoparticles, stabilizers & dispersion medium, **230**: 89

Index

Nanoparticles, surface characteristics, **230**: 90
Nanosilver environmental behavior, influencing factors, **223**: 82
Nanosilver toxicity in vivo, effective concentration (table), **223**: 95
Nanosilver toxicity, causes, **223**: 98
Nanosilver toxicity, influencing factors, **223**: 82
Nanosilver toxicity, ROS mode of action (table), **223**: 85
Nanosilver toxicity, vs. silver ion toxicity, **223**: 89
Nanosilver, applications, **223**: 82
Nanosilver, aquatic organism toxicity, **223**: 83
Nanosilver, aquatic species toxicity, **223**: 93
Nanosilver, biological effects & modes of action, **223**: 81 ff.
Nanosilver, possible modes of action (diag.), **223**: 88
Nanosilver, size-dependent toxicity, **223**: 84
Nanotechnology, defined, **230**: 84
Nanotechnology, for disinfecting water, **222**: 140
Nanotechnology, investment level, **230**: 84
Nanowaste flows, in the environment (diag.), **230**: 88
Natural genetic changes, vs. genetic effects, **227**: 82
Natural rubber, worldwide demand, **227**: 6
Natural sources, cadmium, **229**: 53
Naturally-produced compounds, nanoparticles (table), **230**: 84
Neck exposure, cardiology staff to radiation (diag.), **222**: 81
Nephrotoxic effects on humans, mercury, **226**: 90
Neurodegenerative diseases, mercury, **229**: 10
Neurodegenerative effects, mercury toxicity, **229**: 1 ff.
Neurodegenerative toxicity, defined, **229**: 1
Neuronal degeneration, mercury, **229**: 7
Newborn effects, via addicted mothers (table), **227**: 60–3
Newborn specimen analysis, benefits (diag.), **227**: 68
Ni biomagnification, in aquatic ecosystems, **226**:101 ff.
Nitrification, drinking water, **222**: 145
Nitrogen dioxide exposure, adult asthma cause, **226**: 39
Nivalenol, a B trichothecene mycotoxin, **228**: 110
$NO_2$, air contaminant, **223**: 6
NOEC (no observed effect concentration) for reproduction, E2, **228**: 37

Non-parametric vs. parametric methods, for deriving HC5 values, **230**: 46
Non-target species toxicity, clomazone, **229**: 45
North Pacific gyre, plastic debris accumulation, **227**: 13
$O_3$, air contaminant, **223**: 6
Occupational exposure, bisphenol A, **228**: 60
Ocean bottom, PBM debris density, **227**: 13
Ocean pollutants, PBMs, **227**: 7
Ocean waste, macro PBMs, **227**: 9
Oceanic debris, floating PBM waste, **227**: 10
Oceanic waste, micro PBMs, **227**: 13
Odontocetes, pollutant accumulation, **228**: 2
Organic amendment effect, chromate soil reduction (diag.), **225**: 33
Organic amendment, remediating trace element toxicity (table), **225**: 31
Organic coatings, acid mine drainage control, **226**: 15
Organic materials, flame retardantcy requirements, **222**: 2
Organic matter barriers, to prevent acid mine drainage, **226**: 8
Organic osmotica effects, cadmium, **229**: 65
Organic pollutant effects, on aquatic life, **229**: 28
Organic-rich soil amendments, enhance Cr & Se reduction, **225**: 33
Organics adsorption, biochar, **228**: 86–7
Organism ingestion, PBM fragments, **227**: 25
Organism ingestion, PBMs, **227**: 24
Organochlorine pollutants, and cetaceans, **228**: 2
Organochlorine residues in Guiana dolphins, gender differences, **228**: 9
Organochlorine residues, Guiana dolphin blubber (table), **228**: 7
Organophosphate compounds, flame retardants, **222**: 22
Organophosphorous pesticides, soil column leaching results (table), **221**: 54–57
Organophosphorous pesticides, soil leaching results, **221**: 53
Osprey, PCB toxicity, **230**: 69
Overwintering effects, bioamplification, **227**: 121
Oxalic acid, acid mine drainage control, **226**: 17
Oxidation process, pyrite (diag.), **226**: 4
Oxidation, of arsenite, **224**: 7
Oxidation-reduction reactions, trace elements (table), **225**: 14
Oxidative defense system effects, cadmium, **229**: 66
Oxidative degradation, PBMs, **227**: 20

Oxidative stress in land snails, biomarker of copper pollution, **225**: 119
Oxidative stress role, Bracken fern-induced cancer, **224**: 82
Ozone, drinking water disinfection, **222**: 115
Pakistani fish residues, heavy metal distribution (illus.), **230**: 117
Pakistani fish residues, heavy metals, **230**: 111 ff.
Pakistani marine fish residues, heavy metals, **230**: 124
Pakistani study sites, fish specimens (illus.), **230**: 116
Parametric vs. non-parametric methods, for deriving HC5 values, **230**: 46
Particle (plastic) fragments, POP sorption, **227**: 26
Particle size effects, nanoparticles, **230**: 90
Particulate matter exposure, adult asthma cause, **226**: 39
Particulate matter, air pollutants, **223**: 3
Partition coefficient, diazinon, **223**: 115
Passivation (=microencapsulation), controlling acid mine drainage, **226**: 9
Passivation role, permanganate, **226**: 12
Pathogens of concern, in drinking water, **222**: 129
Pathogens, marine mammal threat, **228**: 2
Pathology, bovine bladder tumor classification, **224**: 66
Pb biomagnification, in aquatic ecosystems, **226**: 101 ff.
Pb residues, in Guiana dolphin kidney & liver (table), **228**: 14
PBDE congener profile, wild lake fish (diag.), **229**: 130
PBDE egg TRVs, for Chinese aquatic avian species (diag.), **229**: 129
PBDE residues, in wild fish from China (diag.), **229**: 132
PBDE residues, in wild fish from Chinese lakes (table), **229**: 131
PBDE toxicity reference values, birds & mammals, **229**: 111 ff.
PBDE TRV derivation, aquatic birds, **229**: 121
PBDE TRV derivation, for aquatic mammals, **229**: 115
PBDE TRV, for aquatic birds, **229**: 125
PBDE TRV, for aquatic mammals, **229**: 119
PBDE TRV, mammals & birds (table), **229**: 127
PBDE TRVs, Chinese aquatic species (diag.), **229**: 129
PBDEs (polybrominated diphenyl ethers) toxicity, to aquatic organisms, **229**: 28

PBDEs (polybrominated diphenylethers), Guiana dolphins, **228**: 10
PBDEs, described, **229**: 112
PBDEs, ecological risk assessment, **229**: 130
PBM (polymer-based material), debris density, ocean bottom, **227**: 13
PBM debris role, alien-species-spread, **227**: 27
PBM degradates, potential effects (diag.), **227**: 25
PBM degradation, in aquatic environments, **227**: 22
PBM degradation, in the natural environment, **227**: 22
PBM degradation, polymer characteristics, **227**: 14
PBM degradation, soil environment, **227**: 22
PBM fragment ingestion, organisms, **227**: 25
PBM pollution, industrial sources, **227**: 8
PBM sink, marine habitat, **227**: 10
PBM waste, agricultural films, **227**: 9
PBMs, abiotic degradation, **227**: 15
PBMs, biotic degradation=biodegradation, **227**: 21
PBMs, categories produced, **227**: 6
PBMs, degradation pathways (diag.), **227**: 15
PBMs, entanglement & ingestion, **227**: 24
PBMs, environmental degradation, **227**: 14
PBMs, environmental effects, **227**: 24
PBMs, environmental occurrence, **227**: 9
PBMs, factors affecting degradation, **227**: 14
PBMs, floating oceanic waste, **227**: 10
PBMs, food chain risks, **227**: 2
PBMs, hydrolytic degradation, **227**: 21
PBMs, in landfills, **227**: 8
PBMs, land-based pollution, **227**: 7
PBMs, manufacturing process, **227**: 3
PBMs, maritime anti-dumping rules, **227**: 7
PBMs, mechanical disintegration, **227**: 21
PBMs, ocean pollutants, **227**: 7
PBMs, oxidative degradation, **227**: 20
PBMs, photodegradation, **227**: 15
PBMs, sewage-related debris, **227**: 8
PBMs, thermal degradation, **227**: 20
PBMs, usage and consumption, **227**: 3
PCB (polychlorinated biphenyls) bird dietary toxicity data, SSD (diag.), **230**: 73
PCB (polychlorinatedbiphenyl) residues, Guiana dolphins, **228**: 6
PCB bird toxicity, SSD (diag.), **230**: 72
PCB isomer-mixture toxicity, to birds (table), **230**: 65–67
PCB mixtures, toxic equivalent conversion factors (table), **230**: 71
PCB residues, in birds (table), **230**: 76

Index

PCB residues, in fish, **230**: 77
PCB residues, in small cetaceans (diag.), **228**: 8
PCB toxicity, American Kestrel, **230**: 70
PCB toxicity, bald eagle, **230**: 69
PCB toxicity, common tern, **230**: 69
PCB toxicity, domestic chicken, **230**: 64
PCB toxicity, double-crested cormorant, **230**: 68
PCB toxicity, Great horned owl, **230**: 70
PCB toxicity, osprey, **230**: 69
PCB TRV-TRG values, derivation methods compared (table), **230**: 75
PCBs, bird toxicity studies, **230**: 64
PCBs, nature and characteristics described, **230**: 60
PCBs, risk assessment to birds, **230**: 76
PCBs, TRVs for aquatic Chinese birds, **230**: 59 ff.
Peat amendment of soil, pesticide mobility effects, **221**: 40
Peat effects, pesticide soil leaching, **221**: 44
Peeling of vegetables & fruits, pesticide residue effect, **229**: 102
Pentaerythritol, intumescent system, **222**: 51
Pentaerythritol, toxic & environmental character, **222**: 52
Pentaerythritol, toxic & physical-chemical properties (table), **222**: 53
Perfluorinated compounds (PFCs), Guiana dolphins, **228**: 10
Performance factors, biochar soil remediation (diag.), **228**: 90
Permanganate, passivation role, **226**: 12
Persistence of HFFRs, classification scheme (table), **222**: 9
Persistence, 9,10-Dihydro-9-oxa-10-phosphaphenanthrene-10-oxide, **222**: 43
Persistence, bisphenol-A bis(diphenylphosphate) (table), **222**: 39
Persistence, bisphenol-A bis(diphenylphosphate), **222**: 38
Persistence, HFFRs (table), **222**: 59
Persistence, HFFRs, **222**: 1 ff.
Persistence, HFFRs, **222**: 58
Persistence, HFFRs, **222**: 6
Persistence, melamine polyphosphate, **222**: 50
Persistence, pentaerythritol, **222**: 52
Persistence, resorcinol bis(diphenylphosphate), **222**: 36
Persistence, triphenylphosphate (table), **222**: 25
Persistence, triphenylphosphate, **222**: 23
Persistent compound consumption, wildlife, **230**: 61
Persistent organic contaminants, in marine mammals, **228**: 5
Persistent organic pollutant air contaminants, **223**: 7
Persistent organic pollutants, bioamplification in wildlife, **227**: 107 ff.
Pesticide absorption-desorption, hysteresis effects, **221**: 21
Pesticide adsorption, biochar, **228**: 86
Pesticide biodegradation effect, plant material amendment of soil, **221**: 41
Pesticide biodegradation rate, method effects, **221**: 24
Pesticide criteria derivation, sediments, **224**: 97 ff.
Pesticide criteria for sediments, methodologies compared (table), **224**: 163
Pesticide degradation, mobility effects, **221**: 19
Pesticide degradation, soil macropore effects, **221**: 33
Pesticide dissipation rate, soil & method effects, **221**: 24
Pesticide leaching effects, colloids & dissolved organic matter, **221**: 7, 8
Pesticide leaching effects, earthworm burrows, **221**: 33
Pesticide leaching, laboratory vs. field studies, **221**: 53
Pesticide leaching, lysimeter (outdoor) results (diag.), **221**: 72
Pesticide leaching, soil adsorption effects, **221**: 26
Pesticide leaching, soils, **221**: 1 ff.
Pesticide leaching, turf thatch effects, **221**: 18
Pesticide loss rate, from soil lysimeters, **221**: 15
Pesticide mobility determination, soil column apparatus (diag.), **221**: 12
Pesticide mobility effect, preferential flow, **221**: 32
Pesticide mobility effects, soil vegetation, **221**: 41
Pesticide mobility in soil, fertilizer effects, **221**: 43
Pesticide mobility measurement, lysimeter advantages, **221**: 16
Pesticide mobility, biodegradation effects, **221**: 23
Pesticide mobility, degradation effects, **221**: 19
Pesticide mobility, elution medium effects (table), **221**: 37
Pesticide mobility, groundwater ubiquity score, **221**: 30
Pesticide mobility, in agriculture, **221**: 3
Pesticide mobility, modeling & simulation methods, **221**: 45

Pesticide mobility, outdoor lysimeter methods, **221**: 14
Pesticide mobility, prospective groundwater studies, **221**: 17
Pesticide mobility, soil amendment & vegetation effects, **221**: 34
Pesticide mobility, soil amendment effects (table), **221**: 37
Pesticide mobility, soil amendment effects, **221**: 36
Pesticide mobility, soil column leaching method, **221**: 11
Pesticide mobility, soil column leaching studies, **221**: 52
Pesticide mobility, soil factor effects, **221**: 31
Pesticide mobility, soil macropore effects, **221**: 3
Pesticide mobility, soil moisture effects, **221**: 31
Pesticide mobility, soil process effects, **221**: 18
Pesticide mobility, soil property effects, **221**: 3
Pesticide mobility, soil thin-layer chromatography, **221**: 10
Pesticide mobility, soil tillage effects (table), **221**: 35
Pesticide mobility, soil tillage effects, **221**: 34
Pesticide mobility, surfactant effects, **221**: 18
Pesticide mobility, urea amendment effects, **221**: 36
Pesticide mobility, vegetation effects (table), **221**: 37
Pesticide residue effect, from boiling or cooking food, **229**: 103
Pesticide residue effect, from food blanching or freezing, **229**: 103
Pesticide residue effect, from food juicing, **229**: 102
Pesticide residue effects, food washing, **229**: 100
Pesticide residue effects, from fruit & vegetable peeling, **229**: 102
Pesticide soil leaching, achieving material balance, **221**: 132
Pesticide soil leaching, peat amendment effects, **221**: 44
Pesticide soil leaching, tannic acid effects, **221**: 44
Pesticide soil mobility, conceptional model (diag.), **221**: 46
Pesticide sorption, soil mobility effects, **221**: 19
Pesticide toxicity, aquatic environments, **224**: 99
Pesticide transport effects, colloid effects, **221**: 7
Pesticide transport in soil, mathematical modeling, **221**: 50
Pesticide transport, dissolved organic matter effects, **221**: 7
Pesticide types, toxicity, **224**: 99
Pesticide types, used in California, **224**: 99
Pesticides (misc.), leaching characteristics, **221**: 74
Pesticides in soil, equilibrium vs. non-equilibrium transport, **221**: 46
Pesticides in soil, facilitated transport, **221**: 42
Pesticides, field leaching, **221**: 16
Pesticides, leaching behavior, **221**: 2
Pesticides, leaching methods, **221**: 10
Pesticides, mobility indices, **221**: 26
Pesticides, structures (appendix), **221**: 77
Pesticides, terrestrial field dissipation studies, **221**: 17
Pesticides, transport processes, **221**: 2
PFCs (perfluorinated compounds), Guiana dolphins, **228**: 10
PGPR (plant growth-promoting rhizobacteria), functions, **223**: 35
PGPR, counter heavy-metal plant stress, **223**: 42
PGPR, heavy metal phytoremediation, **223**: 33ff.
PGPR, in plant heavy-metal remediation, **223**: 43
PGPR, in rhizosphere, **223**: 40
PGPR, plant growth-regulating compounds (table), **223**: 41
pH effects on trace elements, bioaccumulation, **225**: 24
pH effects, on zirconium behavior, **221**: 112
pH effects, soil sorption processes, **221**: 20
Phosphate coating, acid mine drainage control, **226**: 9
Phosphate coating, bonding pattern on pyrite & $FePO_4$ (illus.), **226**: 10
Phosphate coatings, mineral oxidation role, **226**: 10
Photocatalytic breakdown, methomyl (diag.), **222**: 100
Photocatalytic oxidation, phthalates, **224**: 46
Photodegradation, methylmercury, **225**: 19
Photodegradation, PBMs, **227**: 15
Photolysis, clomazone, **229**: 39
Photolysis, diazinon, **223**: 120
Photolysis, fenpropathrin, **225**: 85
Photolysis, methomyl (diag.), **222**: 100
Photolysis, methomyl, **222**: 98
Photolysis, phthalates, **224**: 47
Photolytic breakdown, diazinon pathway in soil (diag.), **223**: 121
Photolytic degradation, clomazone (diag.), **229**: 40

Index 163

Photosynthesis effects, cadmium, **229**: 69
Phthalate esters, abiotic degradation, **224**: 43
Phthalate esters, adverse effects, **224**: 42
Phthalate esters, hydrolysis, **224**: 44
Phthalate esters, in landfills, **224**: 42
Phthalate, structure (illus.), **224**: 41
Phthalates, abiotic conditions in landfills, **224**: 39 ff.
Phthalates, acid & base hydrolysis (diag.), **224**: 45
Phthalates, biodegradation, **224**: 47
Phthalates, chemical behavior, **224**: 39 ff.
Phthalates, chemical-physical properties, **224**: 41
Phthalates, degradation-decomposition, **224**: 43
Phthalates, described, **224**: 39
Phthalates, environmental concentrations, **227**: 33
Phthalates, fugitive nature, **224**: 40
Phthalates, health effects, **224**: 40
Phthalates, hydrolytic transformation (diag.), **224**: 45
Phthalates, photocatalytic oxidation, **224**: 46
Phthalates, photolysis, **224**: 47
Phthalates, synthesis via esterification, **224**: 40
Phylogenetic tools, to assess mutagenicity, **227**: 84
Physical barriers, acid mine drainage, **226**: 4
Physical barriers, to prevent acid mine drainage, **226**: 5
Physical properties, As, Hg & Se (table), **225**: 17
Physical-chemical properties, 9,10-Dihydro-9-oxa-10-phosphaphenanthrene-10-oxide, **222**: 43
Physical-chemical properties, aluminum diethylphosphinate (table), **222**: 46
Physical-chemical properties, aluminum diethylphosphinate, **222**: 45
Physical-chemical properties, ammonium polyphosphate (table), **222**: 15
Physical-chemical properties, bisphenol-A bis(diphenylphosphate) (table), **222**: 39
Physical-chemical properties, HFFRs, **222**: 5
Physical-chemical properties, magnesium hydroxide (table), **222**: 13
Physical-chemical properties, melamine polyphosphate (table), **222**: 49
Physical-chemical properties, melamine polyphosphate, **222**: 49
Physical-chemical properties, pentaerythritol (table), **222**: 53
Physical-chemical properties, pentaerythritol, **222**: 52

Physical-chemical properties, phosphaphenanthrene-10-oxide (table), **222**: 44
Physical-chemical properties, phthalates, **224**: 41
Physical-chemical properties, resorcinol bis(diphenylphosphate), **222**: 35
Physical-chemical properties, triphenylphosphate (table), **222**: 24
Physical-chemical properties, triphenylphosphate, **222**: 23
Physical-chemical properties, zinc borate (table), **222**: 18
Physical-chemical properties, zinc borate, **222**: 17
Physical-chemical properties, zinc hydroxystannate (table), **222**: 19
Physical-chemical properties, zinc stannate (table), **222**: 21
Physicochemical data needs, deriving sediment quality criteria, **224**: 117
Physicochemical properties, clomazone (table), **229**: 37
Physicochemical properties, diazinon (table), **223**: 109
Physicochemical properties, diazinon, **223**: 108
Physicochemical properties, fenpropathrin (table), **225**: 81
Physicochemical properties, fenpropathrin, **225**: 81
Physicochemical properties, methomyl (table), **222**: 95
Physiochemical alterations in plants, cadmium, **229**: 59
Physiochemical effects, cadmium (table), **229**: 56
Phytodegradation, phytoremediation type, **223**: 37
Phytoestrogenic toxin, zearalenone, **228**: 104
Phytoextraction, plant bioremediation, **223**: 35, 36
Phytoextraction, uptake process, **223**: 38
Phytoremediation of heavy metals, rhizobacteria role, **223**: 33ff.
Phytoremediation role, PGPR, **223**: 35
Phytoremediation role, rhizobacteria, **223**: 35
Phytoremediation success, variables, **223**: 35
Phytoremediation, defined, **223**: 36
Phytoremediation, hyperaccumulator plant role, **223**: 37
Phytoremediation, types defined, **223**: 36
Phytostabilization, type of phytoremediation, **223**: 36

Phytostimulation, type of phytoremediation, **223**: 36
Phytotoxic effects, nanoparticles, **230**: 92
Phytotoxicity of nanoparticles, factors affecting potency, **230**: 88
Phytotoxicity, fabricated nanoparticles, **230**: 83 ff.
Phytotoxicity, zirconium, **221**: 107 ff.
Phytovolatilization, type of phytoremediation, **223**: 36
Pilot whale contamination, mercury, **229**: 4
Placenta role, addictive substance biomonitoring, **227**: 64
Placental transport, chemical compounds (diag.), **227**: 67
Plant accumulation, cadmium, **229**: 54
Plant alterations, cadmium, **229**: 59
Plant behavior, cadmium, **229**: 54
Plant detoxification mechanisms, metals, **223**: 39
Plant distribution, zirconium, **221**: 115
Plant effects, clomazone, **229**: 43
Plant growth effects, cadmium (table), **229**: 56
Plant growth inhibition, zirconium, **221**: 117
Plant growth regulators, cadmium effects, **229**: 67
Plant growth-promoting rhizobacteria (PGPR), phytoremediation role, **223**: 35
Plant heavy-metal remediation, PGPR synergy, **223**: 43
Plant levels, copper, **225**: 106
Plant material amendment of soil, pesticide biodegradation effect, **221**: 41
Plant metabolism, fenpropathrin, **225**: 85
Plant metal uptake, cadmium effects, **229**: 64
Plant mode-of-action, clomazone, **229**: 42
Plant penetration, nanoparticles, **230**: 91
Plant pigment effects, cadmium, **229**: 59
Plant root ethylene effect, ACC (diag.), **223**: 45
Plant stress reduction, ACC role, **223**: 44
Plant toxicity effects, reactive oxygen species, **223**: 39
Plant toxicity mechanisms, heavy metals, **223**: 34
Plant toxicity, boron, **225**: 69
Plant toxicity, heavy metals, **223**: 34
Plant toxicity, zirconium, **221**: 107 ff., 116
Plant trait effects, cadmium, **229**: 55
Plant translocation rate, zirconium, **221**: 116
Plant translocation, zirconium, **221**: 114
Plant uptake & transport, metals, **223**: 38
Plant uptake from soil, zirconium, **221**: 114, 115
Plant uptake, zirconium, **221**: 107 ff.
Plant water relations effect, cadmium, **229**: 70

Plant-nanoparticle interactions, mediating factors (diag.), **230**: 89
Plants, and nanoparticles, **230**: 87
Plants, boron deficiency effects, **225**: 62
Plants, boron role, **225**: 61
Plants, containing boron, **225**: 62
Plants, role of heavy metals, **223**: 34
Plant-soil interactions, heavy metals, **223**: 34
Plastic debris accumulation, North Pacific gyre, **227**: 13
Plastic debris effects, Guinea dolphins, **228**: 19
PM10 98$^{th}$ percentile, air pollutant data in Santiago, Chile (table), **223**: 24
PM10 air pollutant data analysis, Santiago, Chile, **223**: 20
PM10 air pollutant data, Santiago, Chile locations (table), **223**: 21
PM10 air pollutant levels, Santiago, Chile monitoring stations (table), **223**: 18
PM10 data analysis, Santiago, Chile, **223**: 17
PM10 level histograms, Santiago, Chile air pollution (diag.), **223**: 19
PM10 levels time series, Santiago, Chile air pollution (diag.), **223**: 19
PNEC (predicted no-effect concentration) of 17-[beta]-estradiol, in Chinese waters, **228**: 29 ff.
PNEC derivation method, E2, **228**: 36
PNEC values for E2, calculated via models (table), **228**: 39
Pollutant accumulation, cetaceans, **228**: 2
Pollutant characteristics, copper, **225**: 97
Pollutant exposures, in fish, **230**: 112
Pollutant threats, marine species, **228**: 2
Pollution assessment, with mammals, **227**: 98
Pollution sources, arsenic, **224**: 5
Polybrominated diphenyl ethers (PBDEs), toxicity reference values, **229**: 111 ff.
Polybrominated diphenylethers, Guiana dolphins, **228**: 10
Polyethylene polyamine, acid mine drainage control, **226**: 16
Polymer additives, described, **227**: 28
Polymer additives, environmental fate, **227**: 28
Polymer additives, environmental implication, **227**: 27
Polymer additives, environmental occurrence, **227**: 30
Polymer additives, environmental residues (table), **227**: 30–2
Polymer additives, hazards described, **227**: 28
Polymer additives, toxicity (table), **227**: 35–39
Polymer additives, toxicity, **227**: 34

Index

Polymer barriers, to prevent acidic mine effluents, **226**: 7
Polymer characteristics, PBM degradation, **227**: 14
Polymer degradation, environmental matrices (table), **227**: 16–19
Polymer types, proportion of consumption, **227**: 6
Polymerase chain reaction, bovine papillomavirus amplification, **224**: 83
Polymer-based materials (PBMs), risks to food chain organisms, **227**: 2
Polymer-based materials, characterized, **227**: 1–2
Polymer-based materials, environmental degradation & effects, **227**: 1 ff.
Polymer-based materials, marine debris component (table), **227**: 11
Polymer-based materials, shoreline debris component (table), **227**: 12
Polymers, flame retardantcy requirements, **222**: 2
Polymers, per capita consumption (table), **227**: 6
Polymers, types and uses (table), **227**: 3–5
POPs (persistent organic pollutants) & bioamplification, modeling, **227**: 124
POPs bioamplification, in wildlife, **227**: 107 ff.
POPs sorption, of PBMs, **227**: 26
POPs sorption, to PBM fragments, **227**: 26
Population effects, of pollution, **227**: 79
Population genetic responses, metal stress, **227**: 84
Population genetics, metal toxicity, **227**: 81
Posttransductional regulation, arsenic forms, **224**: 17
Potable water disinfection, regulation, **222**: 115
Preferential flow, pesticide mobility effect, **221**: 32
Preferential flow, through soil macropores, **221**: 32
Pregnant women, addictive substance abuse (diag.), **227**: 56
Pregnant women, addictive substance effects, **227**: 55 ff.
Pregnant women, addictive substance use, **227**: 58
Pregnant women, implications of addiction, **227**: 58
Prenatal exposure effects, to addictive substances, **227**: 59
Preventing EBH, approaches, **224**: 85
Prevention and control, *Fusarial* toxins, **228**: 110

Prevention measures, acid mine drainage (diag.), **226**: 5
Prevention strategies, acid mine drainage, **226**: 5
Probability plots, PM10 air pollutant data for Santiago, Chile (diags.), **223**: 23, 24
Production volumes, HFFRs, **222**: 6
Proline effects, cadmium, **229**: 62
Properties, clomazone, **229**: 36
Prospective groundwater studies, pesticide mobility, **221**: 17
Protease activity effects, cadmium, **229**: 61
Protein level effects, cadmium, **229**: 61
Protozoan waterborne pathogens, in drinking water, **222**: 133
Ptaquilosid, carcinogenic mechanism (diag.), **224**: 63
Ptaquiloside & similars, structure (illus.), **224**: 59
Ptaquiloside adducts, fern-induced tumors, **224**: 80
Ptaquiloside, analysis method, **224**: 76
Ptaquiloside, cancer types induced, **224**: 62
Ptaquiloside, carcinogen in fern, **224**: 55
Ptaquiloside, carcinogenic mechanism, **224**: 62
Ptaquiloside, content in fern species, **224**: 57, 60
Ptaquiloside, environmental residues, **224**: 74
Ptaquiloside, fern fronds, **224**: 76
Ptaquiloside, genotoxic activity, **224**: 77
Ptaquiloside, genotoxic carcinogen, **224**: 57
Ptaquiloside, human food chain flow (diag.), **224**: 75
Ptaquiloside, human health effects, **224**: 73
Ptaquiloside, in the human food chain, **224**: 73
Ptaquiloside in human food, entry points (diag.), **224**: 74
Ptaquiloside, physiological behavior, **224**: 62
Ptaquiloside, soil & water effects, **224**: 75
Ptaquiloside, soil degradation, **224**: 77
Ptaquiloside, soil leaching, **224**: 76
Ptaquiloside-induced bovine hematuria, environmental effects, **224**: 53 ff.
Ptaquiloside-induced bovine hematuria, human effects, **224**: 53 ff.
Ptaquiloside-induced cancer, tumor modeling, **224**: 78
*Pteridium aquilinum*, EBH cause, **224**: 54
Pulmonata, copper pollution monitoring, **225**: 95 ff.
Punjab Province (Pakistan), fish residues, **230**: 119
Pyrethroid insecticides, residue changes from food processing, **229**: 89 ff.

Pyrethroid pesticides, soil column leaching results (table), **221**: 54–57
Pyrethroid pesticides, soil leaching results, **221**: 53
Pyrethroid residue effects, from food processing methods (table), **229**: 92–99
Pyrite & $FePO_4$, phosphate bonding pattern (illus.), **226**: 10
Pyrite bonding pattern, silica coating (illus.), **226**: 11
Pyrite Fe-oxyhydroxide, coatings cross-section (illus.), **226**: 14
Pyrite, oxidation process (diag.), **226**: 4
Pyrite, silica coatings to prevent acid mine drainage, **226**: 11
Pyrolysis temperature effects, biochar performance, **228**: 87
QSARs (quantitative structure activity relationships), in deriving sediment quality criteria for pesticides, **224**: 130
**R**adiation dose by procedure, cardiology staff (diag.), **222**: 79
Radiation dose for cardiology staff, hands, fingers & wrists (diag.), **222**: 79
Radiation dose for cardiology staff, thyroid/neck region (diag.), **222**: 81
Radiation dose to hands, cardiology staff (diag.), **222**: 82
Radiation dose, cardiology staff eyes & forehead (diags.), **222**: 80
Radiation doses, interventional cardiology staff, **222**: 78
Radiation exposure effects, cardiology staff, **222**: 85
Radiation exposure of cardiology staff, cataract formation, **222**: 84
Radiation exposure of cardiology staff, eye & thyroid gland, **222**: 82
Radiation exposure routes, cardiac catheterization laboratories (illus.), **222**: 87
Radiation exposure to cardiology staff, data sources (diag.), **222**: 76
Radiation exposure, cardiology staff wrists & hands, **222**: 80
Radiation exposure, cardiology staff, **222**: 74
Radiation exposure, vs. radioimaging apparatus, **222**: 77
Radiation safety practices, for cardiology staff, **222**: 85
Radiation, cardiology staff exposure, **222**: 73 ff.
Radioimaging apparatus, vs. radiation exposure, **222**: 77
Rainbow trout toxicity, E2, **228**: 34

Rat acute toxicity, boron, **225**: 67
Rat sub-chronic toxicity, boron, **225**: 68
REACH (Registration, Evaluation, Authorisation and Restriction of Chemical substances) system data, HFFRs, **222**: 5
Reactive oxygen species (ROS), heavy metal induction, **223**: 40
Reactive oxygen species, mercury toxicity, **229**: 8
Redox potential values, trace elements (table), **225**: 14
Redox reactions in the aquatic environment, trace elements (table), **225**: 13
Redox reactions, trace elements in soil (table), **225**: 13
Reduction pathways, arsenite (table), **224**: 8
Reductions vs. oxidations, trace elements in soil, **225**: 15
Reference concentrations, SSD curve construction (table), **230**: 71
Reference concentrations, TRV calculation (table), **230**: 74
Regulating air pollutants, Santiago, Chile, **223**: 4
Regulating water disinfection, USA, **222**: 118
Regulation of water disinfectants, USEPA (table), **222**: 119
Regulations, drinking water disinfection, **222**: 115
Release of gaseous mercury, worldwide sites (table), **226**: 68
Remediation implications, trace elements, **225**: 36
Remediation in soil, trace elements, **225**: 1ff.
Remediation of soils, biochar properties, **228**: 86
Remediation of trace element toxicity, organic amendment (table), **225**: 31
Representative species selection, for TRV setting, **230**: 61
Reproduction effects, bioamplification, **227**: 119
Reproductive effects in birds, methyl mercury, **223**: 62
Reproductive effects in males, bisphenol A derivatives, **228**: 72
Reproductive effects in males, bisphenol A, **228**: 57 ff.
Reproductive effects in males, BPA (table), **228**: 63
Reproductive effects, *in utero* exposure to bisphenol A, **228**: 60–2
Reproductive function in males, BPA action sites (diag.), **228**: 71

Index 167

Reproductive toxicity, SSD for E2 (diag.), **228**: 40
Reptile residues in Mexico, mercury, **226**: 84
Reptile tissue residues, mercury (table), **226**: 85
Residue criteria for methyl mercury, bird protection in China, **223**: 53 ff.
Residues in air, ptaquiloside, **224**: 74
Residues in Bracken fern, ptaquiloside, **224**: 74
Residues in water, ptaquiloside, **224**: 74
Resorcinol bis(diphenylphosphate), flame retardant, **222**: 35
Resorcinol bis(diphenylphosphate), persistence, bioaccumulation & toxicity, **222**: 36
Resorcinol bis(diphenylphosphate), physical-chemical properties, **222**: 35
Resorcinol bis(diphenylphosphate), toxic & physical-chemical properties (table), **222**: 31
Rhizobacteria, in heavy metal phytoremediation, **223**: 33ff.
Rhizobacteria, role in phytoremediation, **223**: 35
Rhizofiltration, type of phytoremediation, **223**: 37
Rhizosphere effects, trace element soil metabolism, **225**: 28
Rhizosphere vs. non-rhizosphere soils, As & Cr metabolic effects (illus.), **225**: 29
Rhizovolatilization, type of phytoremediation, **223**: 36
Risk assessment procedure, metals-metalloids, **230**: 43
Risk assessment role, SSDs, **230**: 37
Risk assessment scheme for nanoparticles, in plants (diag.), **230**: 103
Risk assessment, ambient E2 concentrations in China, **228**: 44
Risk assessment, of 17-[beta]-estradiol, in Chinese waters, **228**: 29 ff.
Risk assessment, PBDEs in Chinese lakes, **229**: 130
Risk assessments, Chinese aquatic species, **230**: 49
Risk management, in drinking water disinfection, **222**: 150, 153
ROS generation, nanosilver toxicity role, **223**: 98
ROS, nanosilver toxicity mechanism, **223**: 82
ROS, nanosilver toxicity mode of action (table), **223**: 85
ROS, plant toxicity effects, **223**: 39
ROS, silver nanoparticle-induced toxicity, **223**: 84
Rural asthma, causes in adults, **226**: 33 ff.
Rural living, asthma morbidity benefit, **226**: 44
Rural vs. urban differences, asthma prevalence (table), **226**: 52
Rural vs. urban environments, differences, **226**: 36
Rural vs. urban residents, asthma effects (table), **226**: 53–55
Rural vs. urban residents, atopy & asthma in adults, **226**: 45
Salt solution washing, during food processing, **229**: 101
Sampling sites, metals-metalloids in a Chinese Lake (illus), **230**: 40
Santiago, Chile air pollutant data, exceedance probabilities & percentiles, **223**: 223
Santiago, Chile air pollutant data, PM10 98th percentile (table), **223**: 24
Santiago, Chile air pollution data, autocorrelation analysis, **223**: 18
Santiago, Chile air pollution, human health effects & guidelines, **223**: 15
Santiago, Chile air pollution, inversion effects, (table), **223**: 14
Santiago, Chile air pollution, PM10 level histograms & boxplots (diag.), **223**: 19
Santiago, Chile air pollution, PM10 levels time series (diag.), **223**: 19
Santiago, Chile air pollution, statistical treatment, **223**: 17
Santiago, Chile locations, PM10 air pollutant data, (table), **223**: 21
Santiago, Chile PM10 levels, boxplots (diag.), (table), **223**: 22
Santiago, Chile, air contamination, **223**: 1ff., 12
Santiago, Chile, air monitoring locations (illus.), **223**: 15
Santiago, Chile, air pollutant characteristics, **223**: 17
Santiago, Chile, air pollutant problems & regulations, **223**: 4
Santiago, Chile, air pollution data, **223**: 16
Santiago, Chile, air pollution monitoring, **223**: 14
Santiago, Chile, PM10 air pollutant levels (table), **223**: 18
Santiago, Chile, PM10 air pollutant data analysis, **223**: 20
Se degradation in soil, reducing conditions, **225**: 20
Secondary poisoning, implication for deriving sediment quality criteria, **224**: 158
Sediment levels, along Mexican coasts, **226**: 68–75

Sediment levels, mercury in Mexico (table), **226**: 69
Sediment pesticide criteria, aquatic life, **224**: 97 ff.
Sediment quality criteria derivation, bioaccumulation & secondary poisoning implication, **224**: 158
Sediment quality criteria derivation, bioavailability considerations, **224**: 135
Sediment quality criteria derivation, calculating criteria, **224**: 131
Sediment quality criteria derivation, data required, **224**: 114
Sediment quality criteria derivation, data sources, **224**: 114
Sediment quality criteria derivation, ecotoxicity data needs, **224**: 120
Sediment quality criteria derivation, ecotoxicity data quality, **224**: 124
Sediment quality criteria derivation, ecotoxicity data quantity needed, **224**: 129
Sediment quality criteria derivation, endpoint choices, **224**: 123
Sediment quality criteria derivation, exposure parameters role, **224**: 132
Sediment quality criteria derivation, interspecies relationship data, **224**: 124
Sediment quality criteria derivation, methodology summary & analysis, **224**: 139–155
Sediment quality criteria derivation, mixtures effect, **224**: 155
Sediment quality criteria derivation, multipathway exposures, **224**: 134
Sediment quality criteria derivation, numeric criteria vs. advisory levels, **224**: 110
Sediment quality criteria derivation, physicochemical data needs, **224**: 117
Sediment quality criteria derivation, QSARs role, **224**: 130
Sediment quality criteria derivation, single- vs. multi-species data, **224**: 122
Sediment quality criteria derivation, species selected (table), **224**: 127
Sediment quality criteria derivation, threatened & endangered species implication, **224**: 157
Sediment quality criteria derivation, value of harmonizing across media, **224**: 160
Sediment quality criteria methodology, survey (table), **224**: 105–6
Sediment quality criteria, biological organization level to protect, **224**: 112
Sediment quality criteria, components (table), **224**: 101
Sediment quality criteria, data analysis methods, **224**: 122
Sediment quality criteria, derivation approaches, **224**: 100
Sediment quality criteria, deriving from numeric criteria vs. advisory levels, **224**: 110
Sediment quality criteria, empirical methods, **224**: 107
Sediment quality criteria, mechanistic approach, **224**: 104
Sediment quality criteria, methodologies compared (table), **224**: 163
Sediment quality criteria, methodology, **224**: 100
Sediment quality criteria, numeric criteria definitions & uses, **224**: 111
Sediment quality criteria, over- & under-protection, **224**: 113
Sediment quality criteria, portion of species to protect, **224**: 113
Sediment quality criteria, protection defined, **224**: 112
Sediment quality criteria, relevant acronyms (table), **224**: 102–4
Sediment quality criteria, spiked-sediment toxicity approach, **224**: 108
Sediment quality criteria, uses & definitions, **224**: 108
Sediment quality guideline values, vs. e-waste contaminants (table), **229**: 25
Sediment quality guidelines, mechanistic approach, **224**: 107
Sediment toxicity testing methods, selected list (table), **224**: 125
Sediment, E2 residues (table), **228**: 45
Sediments characterized, aquatic environments, **224**: 98
Selection forces, genetic structure response, **227**: 80
Selenium characteristics, profile, **225**: 6
Selenium effects, MeHg, **229**: 8
Selenium in soil, uses (table), **225**: 5
Selenium metabolic reduction, influencing factors, **225**: 23
Selenium volatilization, vs. soil water potential (diag.), **225**: 26
Selenium, biomethylation, **225**: 20
Sentinel organisms, use in ecotoxicology studies, **227**: 97
Sewage treatment plant (STP), E2 residues (table), **228**: 45
Sewage-related debris, PBMs, **227**: 8

Shell vs. soft tissues, copper uptake in land snails, **225**: 113
Shoreline debris component, PBMs (table), **227**: 12
Shoreline waste, macro PBMs, **227**: 10
Shoreline waste, micro PBMs, **227**: 13
Shrimp, mercury levels in Mexico (table), **226**: 78
Silica coating, pyrite bonding pattern (illus.), **226**: 11
Silica coatings, for pyrite, **226**: 11
Silver compounds, antibacterial properties, **223**: 89
Silver ions, nanoparticle toxicity cause, **223**: 97, 99
Silver nanoparticle toxicity, influencing factors, **223**: 92
Silver nanoparticle toxicity, ROS, **223**: 84
Silver nanoparticle toxicity, zebrafish, **223**: 94
Silver nanoparticles, bacterial cell-wall interaction, **223**: 90
Silver nanoparticles, daphnid uptake, **223**: 96
Silver nanoparticles, described, **230**: 97
Silver toxicity mechanisms, freshwater fish & daphnids, **223**: 93
Silver toxicity, in vitro exposure, **223**: 83
Silver toxicity, in vivo exposure, **223**: 92
Silver, basal cell function disruption, **223**: 83
S-impregnated activated carbon efficiency, for removing Hg, **230**: 26
S-impregnated activated carbon, for removing Hg, **230**: 2, 4
Sindh Province (Pakistan), fish residues, **230**: 123
Single-walled carbon nanotubes (SWCNTs), description, **230**: 92
Smoking behavior & asthma, urban vs. rural areas, **226**: 37
Snails (Pulmonata), copper pollution monitoring, **225**: 95 ff.
$SO_2$, air contaminant, **223**: 6
Socioeconomic effects, on asthma incidence, **226**: 49
Soil & water effects, ptaquiloside, **224**: 75
Soil adsorption effects, pesticide leaching, **221**: 26
Soil amendment effect, arsenic metabolism (diag.), **225**: 35
Soil amendment effects, pesticide mobility (table), **221**: 37
Soil amendment effects, pesticide mobility, **221**: 34, 36
Soil amendment effects, trace element metabolism, **225**: 30
Soil amendments, chromium immobilization (illus.), **225**: 38
Soil behavior modeling, pesticides, **221**: 45
Soil behavior, arsenic, **225**: 4
Soil behavior, cadmium, **229**: 54
Soil behavior, copper, **225**: 103
Soil behavior, fenpropathrin, **225**: 83
Soil behavior, methomyl, **222**: 95
Soil bioavailability, heavy metals, **223**: 37
Soil bioavailability, trace elements, **225**: 1ff.
Soil bioavailability, zirconium, **221**: 110
Soil biosorption, trace elements (table), **225**: 9
Soil column elution, experimental methods, **221**: 13
Soil column leaching effects, dissolved organic matter, **221**: 8
Soil column leaching studies, pesticide mobility, **221**: 52
Soil column leaching, acetoanilide pesticides (table), **221**: 62–65
Soil column leaching, arylalkanoate pesticides (table), **221**: 62–65
Soil column leaching, carbamate pesticides (table), **221**: 54–57
Soil column leaching, climate effects, **221**: 14
Soil column leaching, diphenyl ether pesticides (table), **221**: 62–65
Soil column leaching, dissolved organic matter effects, **221**: 9
Soil column leaching, measuring pesticide mobility, **221**: 11
Soil column leaching, misc. pesticides (table), **221**: 66–71
Soil column leaching, organophosphorous pesticides (table), **221**: 54–57
Soil column leaching, pyrethroid pesticides (table), **221**: 54–57
Soil column leaching, triazine pesticides (table), **221**: 62–65
Soil column leaching, typical apparatus (diag.), **221**: 12
Soil column leaching, urea & sulfonylurea herbicides (table), **221**: 58–61
Soil column leaching, water-dispersible colloid effects, **221**: 9
Soil columns for pesticide leaching, described, **221**: 12
Soil columns, pesticide leaching, **221**: 1 ff.
Soil degradation of diazinon, influencing factors, **223**: 120
Soil degradation, fenpropathrin, **225**: 84
Soil degradation, methylmercury, **225**: 19
Soil degradation, ptaquiloside, **224**: 77
Soil degradation, Se, **225**: 20

Soil dissipation, diazinon, **223**: 119
Soil effects, pesticide dissipation rate, **221**: 24
Soil environment, PBM degradation, **227**: 22
Soil exposure, copper, **225**: 103
Soil facilitated transport, pesticides, **221**: 42
Soil factor effects, pesticide mobility, **221**: 31
Soil interactions, clomazone, **229**: 37
Soil leaching methods, pesticides, **221**: 10
Soil leaching results, organophosphorous & carbamate pesticides, **221**: 53
Soil leaching results, synthetic pyrethroid insecticides, **221**: 53
Soil leaching, diazinon, **223**: 117
Soil leaching, pesticides, **221**: 1 ff., 2
Soil leaching, ptaquiloside, **224**: 76
Soil macropore effects, pesticide degradation, **221**: 33
Soil macropores, characteristics, **221**: 4
Soil macropores, dye studies, **221**: 4
Soil macropores, earthworm role, **221**: 6
Soil macropores, formation, **221**: 5
Soil macropores, pesticide mobility, **221**: 3
Soil macropores, preferential flow, **221**: 32
Soil metabolism of chromium, influencing factors (diag.), **225**: 24
Soil metabolism of trace elements, rhizosphere effects, **225**: 28
Soil metabolism of trace elements, temperature effects, **225**: 26
Soil metabolism, arsenic, **225**: 19
Soil microbes, biochar aging effect, **228**: 93
Soil microbial degradation, diazinon, **223**: 119
Soil mineralization, diazinon, **223**: 118
Soil mobility of pesticides, conceptional model (diag.), **221**: 46
Soil mobility of pesticides, fertilizer effects, **221**: 43
Soil mobility, groundwater ubiquity score (GUS; diag.), **221**: 30
Soil mobility, zirconium, **221**: 111
Soil moisture & aeration effects, trace element metabolism, **225**: 25
Soil moisture effects, pesticide mobility, **221**: 31
Soil organic matter effects, zirconium mobility, **221**: 112
Soil organic matter, biochar performance effects, **228**: 91
Soil pH, biochar effect, **228**: 90
Soil pH, trace element metabolic effect, **225**: 24
Soil photolysis pathway, diazinon (diag.), **223**: 121
Soil process effects, pesticide mobility, **221**: 18
Soil properties, vs. $K_d$ values, **221**: 22
Soil property effects, pesticide mobility, **221**: 3

Soil remediation, trace elements, **225**: 1ff.
Soil remediation, with biochar (diag.), **228**: 85
Soil remediation, with biochar, **228**: 83 ff.
Soil retention & mobility, zirconium, **221**: 110
Soil sorption effects, pesticide mobility, **221**: 19
Soil sorption processes, pH & clay effects, **221**: 20
Soil sorption, diazinon, **223**: 114, 115
Soil sorption, processes, **221**: 19
Soil texture effects, zirconium behavior, **221**: 112
Soil thin-layer chromatography, pesticide mobility, **221**: 10
Soil tillage effects, pesticide mobility (table), **221**: 35
Soil tillage effects, pesticide mobility, **221**: 34
Soil tillage, macropore disruption, **221**: 31
Soil to plant transfer, zirconium, **221**: 115
Soil transformation, trace elements, **225**: 1ff.
Soil transformation, trace elements, **225**: 3
Soil vegetation, pesticide mobility effects, **221**: 41
Soil water potential, vs. selenium volatilization (diag.), **225**: 26
Soil, trace element sink, **225**: 2
Soil-plant system behavior, zirconium, **221**: 107 ff.
Soil-plant systems, copper transfer, **225**: 108
Soil-plant transfer, zirconium, **221**: 114
Soils, structure & character, **221**: 3
Soils, trace element sources, **225**: 4
Soils, trace element speciation, **225**: 4
Soil-sediment degradation, diazinon, **223**: 114
Solar radiation, drinking water disinfection, **222**: 136
Solute concentration, trace element metabolic effect, **225**: 21
South Africa trends, potable water disinfection, **222**: 124
South Africa, drinking water quality, **222**: 117
Soybean metabolic pathway, clomazone (diag.), **229**: 44
Speciation of trace elements, metabolic effect, **225**: 21
Speciation, zirconium, **221**: 113
Species selected, deriving sediment quality criteria (table), **224**: 127
Species sensitivity distribution (SSD), for E2 toxicity (diag.), **228**: 39
Species sensitivity distribution, methyl mercury avian toxicity (diag.), **223**: 68
Species sensitivity distributions (SSDs), setting WQCs, **230**: 35 ff.

Species sensitivity, to metallic elements, **230**: 46
Species to protect, sediment quality criteria, **224**: 113
Sperm function effects, bisphenol A, **228**: 70
Spermatogenesis effects, of bisphenol A, **228**: 62, 68
Spiked-sediment toxicity approach, sediment quality criteria, **224**: 108
SSD (species sensitivity distribution) bootstrap regression, metal-metalloid toxicity (diag.), **230**: 48
SSD (species sensitivity distribution), E2 toxicity (diag.), **228**: 39
SSD construction, bootstrapping approaches, **230**: 41
SSD construction, parametric approaches, **230**: 40
SSD construction, via different models, **230**: 43
SSD curve construction, reference concentrations (table), **230**: 71
SSD curve fitting, toxicity values (table), **230**: 72
SSD for E2, reproductive toxicity (diag.), **228**: 40
SSD plots, derived via different approaches (diag.), **230**: 45
SSD, method described, **230**: 70
SSD, PCB bird dietary toxicity data (diag.), **230**: 73
SSD, PCB bird toxicity (diag.), **230**: 72
SSDs, derived via different approaches, **230**: 45
SSDs, risk assessment role, **230**: 37
Stabilizers, nanoparticles, **230**: 89
Stack emissions, Hg adsorption variables, **230**: 28
Stack emissions, removing vapor-phase Hg, **230**: 1 ff.
Statistical distribution uses, air pollution data, **223**: 4
Statistical distributions, relation to air pollutant data, **223**: 3, 8
Statistical distributions, Santiago, Chile air contaminants, **223**: 1ff.
Statistical model effect, on HC5 values (diag.), **230**: 44
Statistical model validation, for Santiago, Chile air pollutant data, **223**: 20
Statistical models, air pollution data, **223**: 3
Statistics, relation to adverse human effects, **223**: 24
STPs (sewage treatment plants), E2 residues (table), **228**: 45
Sulfide deposits, acid mine drainage (table), **226**: 3

Sulfide mineral oxidation, acid mine drainage, **226**: 2
Sulfonylurea herbicides, leaching characteristics, **221**: 73
Sulfonylurea herbicides, soil column leaching (table), **221**: 58–61
Sulfur to carbon ratio, $Hg^0$ adsorption effects, **230**: 28
Sulfur-content effects, $Hg^0$ adsorption, **230**: 28
Surface water, E2 residues (table), **228**: 45
Surface-water concentrations, metals-metalloids in China, **230**: 39
Surface-water residues, diazinon, **223**: 123
Surfactant effects, pesticide mobility, **221**: 18
Suspended solid particles, air pollutants, **223**: 7
SWCNTs (single-walled carbon nanotubes), description, **230**: 92
Synthesis methodology, phthalates, **224**: 40
Systematic research reviews, methodology, **230**: 3
Systematic reviews, flow chart process (diag.), **230**: 5
Systematic reviews, process steps (diag.), **230**: 3
*T*-2 and HT-2 toxins, *Fusarium*-produced mycotoxins, **228**: 108
Tannic acid effects, pesticide soil leaching, **221**: 44
Tattoo skin disease, in Guinea dolphins, **228**: 16
Taxa effects, heavy metal uptake (diag.), **226**: 107
Taxonomic group case studies, bioamplification (table), **227**: 116–7
Teleost fish muscle residues, mercury (table), **226**: 82
Temperature effect, diazinon soil mobility, **223**: 118
Temperature effects, $Hg^0$ adsorbed by activated carbon, **230**: 29
Temperature effects, $Hg^0$ adsorption, **230**: 27
Temperature effects, on biochar remediation performance, **228**: 93
Temperature effects, trace element soil metabolism, **225**: 26
Tern (common), PCB toxicity, **230**: 69
Terrestrial ecosystem monitoring, using land snails, **225**: 95 ff.
Terrestrial field dissipation studies, pesticides, **221**: 17
Testicular antioxidant system, bisphenol A effects, **228**: 70
Testicular effects, BPA, **228**: 65
Thermal degradation, PBMs, **227**: 20

Thermal inversion effects, Santiago, Chile air pollution, **223**: 14
Thimerosal in vaccines, ethyl mercury, **229**: 4
Thiol interactions, MeHg, **229**: 7
Thyroid exposure, cardiology staff to radiation (diag.), **222**: 81
Thyroid gland exposure to radiation, cardiology staff, **222**: 82
Time effects, $Hg^0$ adsorbed by activated carbon, **230**: 29
Tissue residue criteria (TRC), for avian wildlife in China, **223**: 53 ff.
Tissue residue guideline (TRG) derivation, methods, **230**: 62
Titanium nanoparticles, described, **230**: 96
Tobacco smoke, adult asthma, **226**: 37
Tolerable daily intake values, for BPA, **228**: 73
Tolerable daily intake values, TRV calculation (table), **230**: 74
Tolerance of land snails, chemical elements, **225**: 97
Tolerance of land snails, to copper forms (table), **225**: 101
Toxic effects survey, nanoparticles (table), **230**: 93–95
Toxic equivalent conversion factors, PCB mixtures (table), **230**: 71
Toxic metals, in e-waste, **229**: 20
Toxic syndromes described, fumonisins, **228**: 106
Toxic trace elements, As, Cr, Hg & Se, **225**: 3
Toxicity among species, copper (table), **225**: 98
Toxicity effect, nanoparticle aggregation, **223**: 94
Toxicity expression methods, copper, **225**: 100
Toxicity influencing factors, silver nanoparticles, **223**: 92
Toxicity mechanism for nanosilver, ROS, **223**: 85
Toxicity mechanisms, heavy metals in plants, **223**: 34
Toxicity of copper, to land snails (table), **225**: 101
Toxicity of HFFRs, classification scheme (table), **222**: 9
Toxicity of Hg, neurodegenerative effects, **229**: 1 ff.
Toxicity of nanosilver, effective in vivo concentration (table), **223**: 95
Toxicity of nanosilver, via ROS induction (table), **223**: 85
Toxicity of PCB isomers-mixtures, to birds (table), **230**: 65-67
Toxicity of silver, in vitro exposure, **223**: 83

Toxicity of silver, in vivo exposure, **223**: 92
Toxicity percentile rank method, for calculating TRVs, **230**: 74
Toxicity reference values (TRV), avian wildlife protection, **223**: 53 ff.
Toxicity reference values (TRV), development process, **229**: 114
Toxicity reference values, aquatic birds in China, **230**: 59 ff.
Toxicity reference values, for PBDEs, **229**: 111 ff.
Toxicity remediation, organic amendments (table), **225**: 31
Toxicity symptoms in humans, boron, **225**: 67
Toxicity symptoms, mercury, **229**: 6
Toxicity thresholds for avian species, methyl mercury (diag.), **223**: 65
Toxicity thresholds for DE-71, wildlife (diag.), **229**: 127
Toxicity thresholds for methyl mercury, species sensitivity curve fitting (table), **223**: 67
Toxicity to aquatic organisms, methomyl (table), **222**: 104
Toxicity to birds & mammals, methomyl, **222**: 104
Toxicity values, SSD curve fitting (table), **230**: 72
Toxicity, 9,10-Dihydro-9-oxa-10-phosphaphenanthrene-10-oxide, **222**: 43
Toxicity, aluminum diethylphosphinate (table), **222**: 46
Toxicity, aluminum diethylphosphinate, **222**: 48
Toxicity, aluminum trihydroxide, **222**: 10
Toxicity, ammonium polyphosphate (table), **222**: 15
Toxicity, ammonium polyphosphate, **222**: 16
Toxicity, arsenic, **224**: 2
Toxicity, bisphenol-A bis(diphenylphosphate) (table), **222**: 41
Toxicity, bisphenol-A bis(diphenylphosphate) (table), **222**: 42
Toxicity, boron, **225**: 66
Toxicity, heavy metals in plants, **223**: 34
Toxicity, HFFRs (table), **222**: 59
Toxicity, HFFRs, **222**: 1 ff.
Toxicity, HFFRs, **222**: 58
Toxicity, HFFRs, **222**: 7
Toxicity, magnesium hydroxide (table), **222**: 13
Toxicity, magnesium hydroxide, **222**: 12
Toxicity, melamine polyphosphate (table), **222**: 50
Toxicity, melamine polyphosphate, **222**: 51
Toxicity, mercury vapor, **229**: 6

Index 173

Toxicity, mercury, **229**: 5
Toxicity, methomyl to aquatic species, **222**: 103
Toxicity, methomyl, **222**: 102
Toxicity, methomyl, **222**: 93 ff.
Toxicity, methomyl, **222**: 94
Toxicity, pentaerythritol (table), **222**: 54
Toxicity, pentaerythritol, **222**: 52
Toxicity, pesticide types, **224**: 99
Toxicity, pesticides in aquatic environments, **224**: 99
Toxicity, phosphaphenanthrene-10-oxide (table), **222**: 44
Toxicity, polymer additives (table), **227**: 35–39
Toxicity, polymer additives, **227**: 34
Toxicity, resorcinol bis(diphenylphosphate) (table), **222**: 33
Toxicity, resorcinol bis(diphenylphosphate), **222**: 36
Toxicity, triphenylphosphate (table), **222**: 27
Toxicity, triphenylphosphate, **222**: 30
Toxicity, zinc borate (table), **222**: 18
Toxicity, zinc borate, **222**: 17
Toxicity, zinc hydroxystannate, **222**: 20
Toxicity, zinc stannate (table), **222**: 21
Toxicology, clomazone, **229**: 35 ff., 42
Toxin sensing, microbes, **224**: 26
Toxoplasmosis, in marine mammals, **228**: 15
Trace element biosorption, by microbes, **225**: 11
Trace element detoxification, methylation & demethylation, **225**: 16
Trace element immobilization, microbial aspects (diag.), **225**: 36
Trace element interactions, soil microbes, **225**: 2
Trace element metabolic effect, soil pH, **225**: 24
Trace element metabolism, microbial activity, **225**: 21
Trace element metabolism, soil amendment effects, **225**: 30
Trace element metabolism, soil moisture & aeration effects, **225**: 25
Trace element metabolism, solute concentration effect, **225**: 21
Trace element metabolism, speciation effect, **225**: 21
Trace element sink, soil, **225**: 2
Trace element soil metabolism, rhizosphere effects, **225**: 28
Trace element soil metabolism, temperature effects, **225**: 26
Trace element sources, soils & sediments, **225**: 4

Trace element speciation, soils & sediments, **225**: 4
Trace element toxicity, organic amendment remediation (table), **225**: 31
Trace elements in soil, microbial biomethylation, **225**: 17
Trace elements, bioaccumulation & remediation effects, **225**: 36
Trace elements, bioaccumulation, **225**: 7
Trace elements, biosorption (table), **225**: 9
Trace elements, defined, **225**: 2
Trace elements, methylation & demethylation in soil, **225**: 18
Trace elements, microbial redox reactions, **225**: 12
Trace elements, microbial transformation processes (diag.), **225**: 8
Trace elements, microbial transformation processes, **225**: 20
Trace elements, oxidation-reduction reactions (table), **225**: 14
Trace elements, pH effects on bioaccumulation, **225**: 24
Trace elements, physical properties (table), **225**: 17
Trace elements, redox reactions (table), **225**: 13
Trace elements, remediation and bioavailability, **225**: 1ff.
Trace elements, soil microbial transformation, **225**: 1ff.
Trace elements, soil reductions vs. oxidations, **225**: 15
Transport model for xenobiotics, mother to child (diag.), **227**: 65
TRC (tissue residue criteria), for avian wildlife in China, **223**: 53 ff.
TRC, for Chinese bird protection, **223**: 53 ff.
TRCs for selected avian species in China, methyl mercury (table), **223**: 67
TRG (tissue residue guideline) derivation, methods, **230**: 62
TRG calculation, methods compared for PCBs (table), **230**: 75
TRG derivation, method, **230**: 70
Triazine pesticides, leaching characteristics, **221**: 73
Triazine pesticides, soil column leaching (table), **221**: 62–65
Trichothecene mycotoxin, diacetoxyscirpenol, **228**: 109
Trichothecene mycotoxin, nivalenol, **228**: 110
Trichothecenes, *Fusarium*-produced mycotoxin, **228**: 108
Triphenylphosphate, flame retardant, **222**: 22

Triphenylphosphate, toxic, environmental & physical-chemical properties, **222**: 22–35
Triphenylphosphate, toxic, environmental & physical-chemical properties (table), **222**: 23–29
Trophic transfer data analysis, method described, **226**: 103
Trophic transfer factor estimates, from biokinetic modeling, **226**: 104
Trophic transfer factors, biokinetic heavy metal data (diag.), **226**: 110
Trophic transfer factors, defined, **226**: 102
Trophic transfer factors, dietary uptake of heavy metals (diag.), **226**: 107
Trophic transfer in freshwater, heavy metals (diag.), **226**: 109
Trophic transfer of accumulator species, biokinetic data (diag.), **226**: 112
Trophic transfer, defined, **226**: 102
TRV & TRC derivation, avian species in China, **223**: 66
TRV & TRC values for methyl mercury, reasonableness, **223**: 68
TRV (toxicity reference value) calculation, methods compared for PCBs (table), **230**: 75
TRV (toxicity reference value), for avian wildlife in China, **223**: 53 ff.
TRV (toxicity reference values), development process, **229**: 114
TRV and TRC values, derivation methods, **223**: 57
TRV calculation, reference concentrations (table), **230**: 74
TRV calculation, tolerable daily intake values (table), **230**: 74
TRV calculation, via the toxicity percentile rank method, **230**: 74
TRV derivation, method, **230**: 70
TRV derivation, methods, **230**: 62
TRV setting, role in protecting wildlife, **223**: 55
TRV, for avian wildlife in China, **223**: 53 ff.
TRVs & TRCs, uncertainty evaluation, **223**: 72
TRVs for selected avian species in China, methyl mercury (table), **223**: 67
TRVs, aquatic birds in China, **230**: 59 ff.
TRV-setting, selecting representative species, **230**: 61
Tumor biomarkers, expression, **224**: 80
Tumor induction & progression, role for bovine papillomavirus, **224**: 64

Tumor modeling, ptaquiloside-induced cancer, **224**: 78
Tumorous disease of cattle, enzootic bovine hematuria, **224**: 53 ff.
Turf thatch effects, pesticide leaching, **221**: 18
Ultrasonic disinfection, of drinking water, **222**: 139
Ultrasonography, cancer diagnostic aid, **224**: 82
United States registered uses, fenpropathrin, **225**: 78
Universal PCR (polymerase chain reaction) primers, fungi (table), **228**:126
Urban asthma, causes in adults, **226**: 33 ff.
Urban vs. agricultural uses in California, diazinon (table), **223**: 113
Urban vs. rural areas, adult asthma prevalence & morbidity, **226**: 36
Urban vs. rural areas, endotoxin levels, **226**: 42
Urban vs. rural areas, fungal allergen levels, **226**: 42
Urban vs. rural differences, asthma prevalence (table), **226**: 52
Urban vs. rural environments, differences, **226**: 36
Urban vs. rural residents, asthma effects (table), **226**: 53–55
Urban vs. rural residents, atopy & asthma in adults, **226**: 45
Urea amendment effects, pesticide mobility, **221**: 36
Urea herbicides, characteristics, **221**: 73
Urea herbicides, soil column leaching (table), **221**: 58–61
USA trends, potable water disinfection, **222**: 124
USA, water disinfectant regulations, **222**: 118
Usage, PBMs, **227**: 3
USEPA, maximum residual disinfectant levels (table), **222**: 118
USEPA, regulating water disinfectants (table), **222**: 119
Uses, of major polymer types (table), **227**: 3–5
**V**apor-phase Hg removal, by sulfur-impregnated activated carbon, **230**: 1 ff.
Vapor-phase Hg removal, stack emissions, **230**: 1 ff.
Vegetable processing, pyrethroid insecticide residue effect, **229**: 89 ff.
Vegetation effects, pesticide mobility (table), **221**: 37
Vegetation effects, pesticide mobility, **221**: 34
Vertebrates in Mexico, mercury residues, **226**: 79
Vestimentiferan tube worms in Mexico, mercury residues (table), **226**: 88

Index

Viral waterborne pathogens, in drinking water, **222**: 133
Virus diseases, Guinea dolphins, **228**: 17
Volatile organic compounds, air contaminants, **223**: 6
Volatilization loss, diazinon, **223**: 127
Volatilization, chlomazone, **229**: 39
Water & soil effects, ptaquiloside, **224**: 75
Water behavior, clomazone, **229**: 38
Water behavior, fenpropathrin, **225**: 83
Water behavior, methomyl, **222**: 96
Water contamination by fungi, control, **228**:131
Water cover, acid mine drainage, **226**: 6
Water disinfectant regulations, USA, **222**: 118
Water disinfection trends, Australia & Europe, **222**: 127
Water disinfection trends, Canada, **222**: 126
Water disinfection trends, South Africa, **222**: 124
Water disinfection trends, USA, **222**: 124
Water disinfection, benchmarking processes, **222**: 154
Water disinfection, by-products present (table), **222**: 142
Water disinfection, global trends, **222**: 124
Water disinfection, history, **222**: 112, 113
Water disinfection, microbial risk assessment, **222**: 153
Water disinfection, practice survey (table), **222**: 125
Water disinfection, successes & challenges, **222**: 111 ff.
Water monitoring data for California, diazinon (table), **223**: 124
Water quality criteria (WQC), metals & metalloids in China, **230**: 35 ff.
Water quality criteria, diazinon (table), **223**: 133
Water quality criteria, diazinon**223**: 133
Water residues of mercury, in Mexico (table), **226**: 76
Water residues, diazinon, **223**: 122
Water safety planning, process (diag.), **222**: 152
Water, sources & types of fungi, **228**:123
Water-dispersible colloids, pesticide leaching effects, **221**: 7
Watersheds in California, diazinon residues (table), **223**: 124
Weight-loss link, bioamplification, **227**: 111
White ibis, methyl mercury toxicity, **223**: 63
Wild fish from Chinese lakes, PBDE residues (diag.), **229**: 132
Wild fish in Chinese lakes, PBDE congener profile (diag.), **229**: 130
Wild fish residues in China, PBDE (table), **229**: 131
Wildlife contaminant, methyl mercury, **223**: 55
Wildlife protection, TRV role, **223**: 55
Wildlife, persistent compound consumption, **230**: 61
Wildlife, POPs bioamplification, **227**: 107 ff.
Wind inversion effects, Santiago, Chile air pollution, **223**: 14
World mine production trend, zirconium (diag.), **221**: 110
Worldwide contaminant, mercury, **226**: 65
Worldwide demand, natural rubber, **227**: 6
WQC (water quality criteria), metals & metalloids in China, **230**: 35 ff.
WQC method-setting, in China, **230**: 35 ff.
WQC standards, need in China, **230**: 36
WQC, defined & described, **230**: 36
WQC-setting, for aquatic Chinese species, **230**: 38
WQC-setting, methods & approaches, **230**: 38
WQC-setting, methods discussed, **230**: 51
Wrists, dose to cardiology staff (diag.), **222**: 79
Xenobiotic transport model, mother to child (diag.), **227**: 65
Xenobiotics, gene pool effects, **227**: 82
Yellow perch bioamplification factors, modeling (table), **227**: 129
Yellow perch simulation, bioamplification model parameters (table), **227**: 138–9
Yellow perch simulation, bioamplification, **227**: 137
Yellow perch, bioamplification modeling (diags.), **227**: 127
Zearalenone, phytoestrogenic toxin, **228**: 105
Zebrafish, silver nanoparticle toxicity, **223**: 94
Zinc borate, flame retardant, **222**: 16
Zinc borate, toxicity & physical-chemical properties (table), **222**: 18
Zinc borate, toxicity & physical-chemical properties, **222**: 17
Zinc hydroxystannate, bioaccumulation & toxicity, **222**: 20
Zinc hydroxystannate, flame retardant, **222**: 18
Zinc hydroxystannate, physical-chemical properties, **222**: 18
Zinc hydroxystannate, toxicity & physical-chemical properties (table), **222**: 19
Zinc stannate, bioaccumulation, **222**: 22
Zinc stannate, flame retardant, **222**: 21
Zinc stannate, toxicity & physical-chemical properties (table), **222**: 21
Zinc stannate, toxicity & physical-chemical properties, **222**: 21

Zink nanoparticles, described, **230**: 96
Zirconium behavior, pH & soil texture effects, **221**: 112
Zirconium behavior, soil-plant systems, **221**: 107 ff.
Zirconium isotopes, described, **221**: 109, 110
Zirconium minerals, described, **221**: 109
Zirconium mobility, soil organic matter effects, **221**: 112
Zirconium release, climate effects, **221**: 112
Zirconium speciation, biogeochemical behavior effects, **221**: 113
Zirconium toxicity, plant defenses, **221**: 117
Zirconium, commercial uses, **221**: 109
Zirconium, crustal abundance, **221**: 113
Zirconium, description, **221**: 108
Zirconium, factors affecting mobility & bioavailability, **221**: 112
Zirconium, mineral structure, **221**: 111
Zirconium, minerals & isotopes, **221**: 108
Zirconium, organism toxicity, **221**: 116
Zirconium, plant growth inhibition, **221**: 117
Zirconium, plant translocation rate, **221**: 116
Zirconium, plant translocation, **221**: 114
Zirconium, plant uptake & distribution, **221**: 115
Zirconium, plant uptake from soil, **221**: 114
Zirconium, soil bioavailability, **221**: 110
Zirconium, soil mobility, **221**: 111
Zirconium, soil retention & mobility, **221**: 110
Zirconium, soil to plant transfer, **221**: 115
Zirconium, sources & abundance, **221**: 108
Zirconium, speciation, **221**: 113
Zirconium, toxicity, **221**: 116
Zirconium, worldwide production trend (diag.), **221**: 110
Zn HC5 values, analysis approach effects (diag.), **230**: 47